Springer Theses

Recognizing Outstanding Ph.D. Research

For further volumes:
http://www.springer.com/series/8790

Aims and Scope

The series "Springer Theses" brings together a selection of the very best Ph.D. theses from around the world and across the physical sciences. Nominated and endorsed by two recognized specialists, each published volume has been selected for its scientific excellence and the high impact of its contents for the pertinent field of research. For greater accessibility to non-specialists, the published versions include an extended introduction, as well as a foreword by the student's supervisor explaining the special relevance of the work for the field. As a whole, the series will provide a valuable resource both for newcomers to the research fields described, and for other scientists seeking detailed background information on special questions. Finally, it provides an accredited documentation of the valuable contributions made by today's younger generation of scientists.

Theses are accepted into the series by invited nomination only and must fulfill all of the following criteria

- They must be written in good English.
- The topic should fall within the confines of Chemistry, Physics, Earth Sciences, Engineering and related interdisciplinary fields such as Materials, Nanoscience, Chemical Engineering, Complex Systems and Biophysics.
- The work reported in the thesis must represent a significant scientific advance.
- If the thesis includes previously published material, permission to reproduce this must be gained from the respective copyright holder.
- They must have been examined and passed during the 12 months prior to nomination.
- Each thesis should include a foreword by the supervisor outlining the significance of its content.
- The theses should have a clearly defined structure including an introduction accessible to scientists not expert in that particular field.

Nick Bahrami

Evaluating Factors Controlling Damage and Productivity in Tight Gas Reservoirs

Doctoral Thesis accepted by
Curtin University, Australia

Springer

Author
Dr. Nick Bahrami
Petroleum Engineering
Curtin University
Perth, WA
Australia

Supervisor
Assoc. Prof. Reza Rezaee
Petroleum Engineering
Curtin University
Perth, WA
Australia

ISSN 2190-5053 ISSN 2190-5061 (electronic)
ISBN 978-3-319-02480-6 ISBN 978-3-319-02481-3 (eBook)
DOI 10.1007/978-3-319-02481-3
Springer Cham Heidelberg New York Dordrecht London

Library of Congress Control Number: 2013949482

Printed on acid-free paper

Springer is part of Springer Science+Business Media (www.springer.com)

Parts of this thesis have been published in the following journal articles:

(1) Bahrami, H., Rezaee. R., Clennell, B., 2012. Water Blocking Damage in Hydraulically Fractured Tight Sand Gas Reservoirs, An Example from Perth Basin, Western Australia. Journal of Petroleum Science and Engineering

(2) Bahrami H., Rezaee R., Hossain M, 2012. Characterizing Natural Fractures Productivity in Tight Gas Reservoirs, Journal of Petroleum Exploration and Production Technology

(3) Bahrami H., Jayan. V., Rezaee. R., Hossain, M. M., 2012. Welltest analysis of hydraulically fractured tight gas reservoirs: An Example from Perth Basin, Western Australia. APPEA Journal

(4) Bahrami H., Rezaee. R., Nazhat D., Ostojic J., 2011. Evaluation of damage mechanisms and skin factor in tight gas reservoirs. APPEA Journal

(5) Murickan G., Bahrami H., Rezaee R., Saeedi A., Mitchel P. A .T., 2012. Using relative permeability curves to evaluate phase trapping damage caused by water-based and oil-based drilling fluids in tight gas reservoirs. APPEA Journal

(6) Ostojic, J., Rezaee, R., and Bahrami, H., 2011. Hydraulic fracture productivity performance in tight gas sands—A Numerical Simulation Approach. Journal of Petroleum Science and Engineering

(7) Bahrami H., Rezaee. R., Rasouli V., Hosseinian A., 2010. Liquid loading in wellbore and its effect on well clean-up period and well productivity in tight gas reservoirs, APPEA Journal, Brisbane, Australia

Supervisor's Foreword

The study brings new insights regarding tight gas reservoirs characterization for dynamic parameters, integrates the reservoir and production engineering of tight gas reservoirs with the other aspects such as petrophysics and geomechanics, and quantifies the effect of different parameters on tight gas sand reservoirs productivity. Core analysis experiments are conducted and the data are studied to determine the relative permeability and capillary pressure curves and also to understand the effect of invasion of different drilling fluids into the tight reservoirs on the damage mechanisms and well productivity. Furthermore, new methods of dynamic data analysis are proposed for more reliable estimation of the effective reservoir permeability for the tight formations. The thesis results have already inspired further work in progress on tight gas reservoirs characterization, damage mitigation in tight gas reservoirs using thermal methods such as down-hole heating, and optimizing the development plans of Whicher Range tight gas field reservoir in Western Australia.

Perth, August 2013

Reza Rezaee

Preface

The Ph.D. study was started in July 2009 and several papers were reviewed on different aspects of tight gas reservoirs. The objectives of this study were to review and evaluate the factors in tight gas reservoirs that have significant influence on formation damage and well productivity, based on field data analysis, laboratory core analysis experiments, analytical approaches, and numerical simulation. In the early stage of the study, typical tight gas data were gathered from the reviewed papers, to build tight gas reservoirs simulation models at core scale, well scale and reservoir scale. Then after gathering actual laboratory and field data, the simulation models were updated to get more realistic results. The simulation models were run using industry-standard softwares that have the advantage of a high degree of validation in real-world situations. The core flooding experiments in this research study were performed using the lab facilities of Petroleum Engineering Department at Curtin University, which have been designed, developed and setup by Dr. Ali Saeedi [1].

A summary of my research over the first two years of Ph.D. studies was presented in the 2011 SPE European Formation Damage Conference, and the published paper went on the list of top 10 downloaded papers from the SPE e-library, which showed interest of the industry in results of the tight gas damage and productivity evaluation study. During the course of this research, several technical papers were published in peer-reviewed journals, all of which were relevant to the work carried out in this research. Every paper was peer-reviewed by at least two expert reviewers and their comments were applied to improve this work. These papers cover the main aspects of this research. Consequently, this Ph.D. thesis is presented based on the published papers which are explained briefly in the body of the thesis report and in more detailed in the journal papers published based on outcomes of this research work. Some sections that have not been published as journal papers, they are explained in more details in the thesis report.

Concerning the written thesis, Chap. 1 presents a brief introduction about the problems associated with tight gas reservoirs and a review of past studies conducted by other researchers. The objectives and significance of this research are outlined. In Chap. 2, determination of the effective permeability of tight formations is discussed and the new techniques are proposed. In Chap. 3, reservoir simulation studies for different types of well and tight reservoirs are illustrated. In Chap. 4, the tight gas field data are analyzed and the well productivity issues in this field are made clear. Finally, a summary of this work is presented in Chap. 5, followed by conclusions and recommendations.

Reference

1. Saeedi Ali (2012) Experimental study of multi-phase flow in porous media during CO_2 geo-sequestration processes, Springer Theses.

Acknowledgments

This work could have not been completed without the help and support of many individuals during these three years, especially my beloved parents who always supported me in all stages of my life. I would like to express my deep and sincere gratitude to my Supervisor Associate Prof. Reza Rezaee for kindly supporting me with his great experience and knowledge, offering invaluable assistance and guidance throughout my studies, and providing a clear road map to complete my thesis work. I am greatly thankful to Dr. Ben Clennell and Dr. Mofazzal Hossain and Dr. Ali Saeedi for detailed and constructive technical guide, advices, help, and valuable feedback that significantly improved quality of my Ph.D. research studies. I wish to give my sincere thanks to Prof. Brian Evans, head of the Petroleum Engineering Department at Curtin University, and Associate Prof. Vamegh Rasouli for their kind support regarding my Ph.D. research studies and giving valuable feedback about my progress. I wish to extend my warmest thanks to all my colleagues and postgraduate students in the Petroleum Engineering Department who have helped me with my work. I would like to thank Ian Lusted and David Pena (Aurora Oil and Gas) for their kind supports and practical advices regarding my thesis on fractured tight reservoirs characterisation. I appreciate KAPPA Engineering, Computer Modelling Group of Canada, and Strategy Central for their generous support by providing free licenses of their software to Curtin University's Petroleum Engineering Department.

Nick Bahrami

Contents

Abbreviations

μ	Viscosity
φ	Porosity
$\sigma_{h,max}$	Maximum horizontal stress
$\sigma_{h,min}$	Minimum horizontal stress
σ_v	Vertical stress
a_f	Fracture spacing
b_f	Aperture of natural fracture
B	Formation volume factor
C	Compressibility
GWC	Gas water contact
h	Thickness
ID	Internal diameter
K	Permeability
K_d	Damaged zone permeability
K_f	Fracture permeability
K_r	Relative permeability
m	Slope of pressure derivative on welltest analysis plots
MMSCFD	Million standard cubic feet per day
n	Number (for example, number of hydraulic fractures)
OB	Overbalanced
P	Pressure
P'	Pressure derivative
P''	The second derivative of transient pressure
P_c	Capillary pressure
Q	Flow rate
Q_g	Gas production rate
Q_d	Dimensionless production rate
S	Skin
σ	In-situ stress
S_{gc}	Critical gas saturation
S_{gi}	Initial gas saturation
$S_{w,\ connate}$	Connate Water Saturation
S_{wi}	Initial Water Saturation
t	Time

t_p	Production time
T	Temperature
UB	Underbalanced
W_f	Aperture of hydraulic fracture
X	Direction in x direction (horizontal)
X_f	Hydraulic fracture half length size
Y	Direction in Y direction (horizontal)
Z	Direction in Z direction (vertical)

Chapter 1
Introduction to Tight Gas Reservoirs

1.1 Introduction

Tight gas reservoirs represent a significant portion of natural gas reservoirs worldwide. Tight gas sand is generally characterized as a formation with effective permeability less than 0.1 md [1]. Production at economical rates from tight gas reservoirs in general is very challenging not only due to the very low intrinsic permeability but also as a consequence of several different forms of formation damage that can occur during drilling, completion, stimulation, and production operations [2, 3]. Tight gas reservoirs generally do not flow gas to surface at commercial rates, unless the well is efficiently completed and stimulated using advanced technologies [4, 5].

1.2 Tight Gas Reservoirs Characteristics

Matrix permeability in tight sand formations may be very low due to the depositional processes, or because of the post-depositional diagenetic events [6]. If the tight gas reservoir is naturally fractured, then the gas flow is mainly controlled by the open undamaged natural fractures that are connected to the wellbore [7].

The rock matrix may primarily be composed of micro-pores where average pore throat aperture is very small, causing tremendous amounts of potential capillary pressure energy suction. In tight formations that are water-wet in nature, the capillary forces cause liquid to be imbibed and held in the capillary pores. This causes the critical water saturation and irreducible water saturation to be high in the tight formations [8, 9]. Initial water saturation (S_{wi}) in tight gas reservoirs might vary depending on the timing of gas migration. A tight gas reservoir may have normal initial water saturation ($S_{wi} \sim S_{wc}$) or in some cases sub-normal ($S_{wi} \ll S_{wc}$) due to water phase vaporization into the gas phase as shown in Fig. 1.1. A sub-normal S_{wi} provides relatively a higher effective permeability for gas phase, close to absolute permeability. The initial water saturation might also be more than critical water saturation if the hydrocarbon trap is created during or after

N. Bahrami, *Evaluating Factors Controlling Damage and Productivity in Tight Gas Reservoirs*, Springer Theses, DOI: 10.1007/978-3-319-02481-3_1,

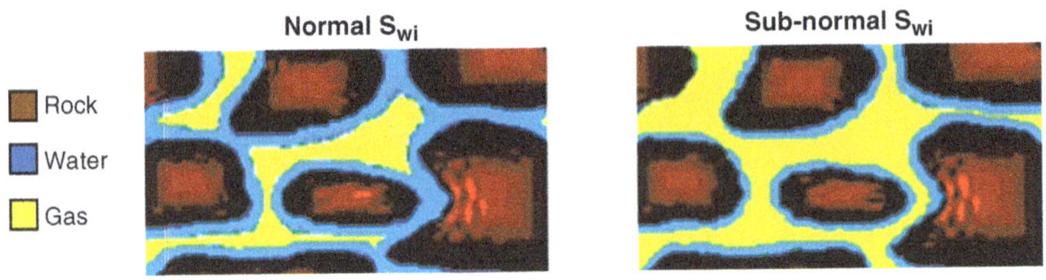

Fig. 1.1 Normal and sub-normal S_{wi} in tight gas reservoirs

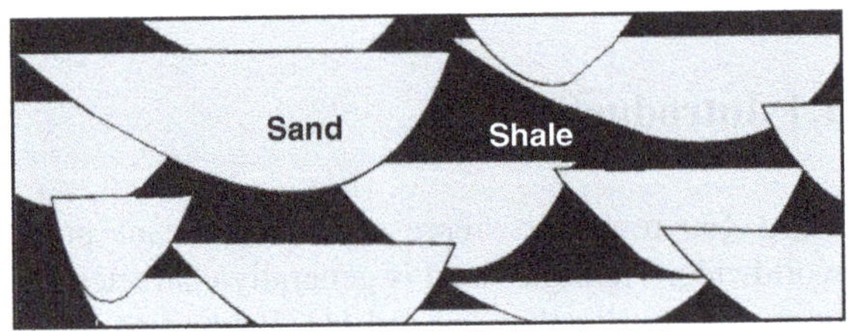

Fig. 1.2 Typical sand lenses in tight gas reservoirs

the gas migration time. In the case of high initial water saturation, relative permeability to gas may be very low [6, 10].

The reservoir geometry of tight gas reservoirs depends on their deposition environment: they normally consist of numerous reservoir layers/lenses, which are discontinuous both vertically and laterally in a thick complex sedimentary system, and separated by non-reservoir shales. The stacks of isolated lenses of sand bodies may vary in characteristics, shape and volume as shown in Fig. 1.2 [11]. The recoverable gas in place in tight gas reservoirs is mainly controlled by the sand lens width, and the effect of sand lens length is not very significant [12]. Horizontal deviated well drilling can help intersecting as many of the sand lenses as possible, and effectively increase lateral reservoir exposure to wellbore [2].

Tight gas reservoirs productivity may also be affected by in situ stresses as they can control hydraulic fractures propagation, reservoir flow regimes, permeability anisotropy, wellbore instability and long term production performance. Tight gas sandstones are typically very stiff rocks capable of supporting high, or even extreme deviatoric stresses.

Tight sand formations commonly have wellbore instability issues during drilling, which causes large wellbore breakouts and washouts across the tight sand intervals. The wellbore instability issues in tight formations can be reduced by drilling the well in the minimum stress direction [13].

In tight sand reservoirs, understanding of the relative magnitude of in situ stresses and their direction, and their relationship with permeability are essential for tight gas development [14, 15]. Tight gas reservoirs are normally heterogeneous and anisotropic in nature, where permeability is a direction dependent property. The permeability anisotropy may be further enhanced by the pattern of earlier geological deformation and be amplified by in situ stresses [16].

In cased-hole perforated wells completed in tight sand, well productivity may significantly be controlled by perforation parameters. Perforation performance depends on factors such as length of down-hole penetration, shot phasing, and shot density. Deep penetration, at least 50 % beyond the damage thickness, is needed to effectively connect wellbore to undamaged rock. Perforation efficiency in tight gas reservoirs is affected by the high rock strength that makes penetration of perforation jet to be significantly reduced compared with equivalent sandstone of higher porosity. Using deep penetrating perforation charges run with shock absorbers can mitigate damage to perforation tunnels and reduce the skin factor [17].

1.3 Damage Mechanisms in Tight Gas Reservoirs

Tight gas reservoirs can be subject to different damage mechanisms during well drilling, completion, stimulation and fracturing, such as mechanical damage to formation rock, plugging of natural fractures by invasion of mud solid particles, permeability reduction around wellbore mainly as a result of filtrate invasion, clay swelling and liquid phase trapping [18]. Materials such as mud filtrate, cement slurry, or clay particles may enter the open pores of the formation and reduce permeability around the wellbore as well [14]. The solids may penetrate only a short distance into the rock matrix and cause only a shallow mechanically damaged zone. However the damage to natural fractures and open permeable conduits can be severe, as drilling fluids invasion mostly occurs through the natural fractures [19, 20] .

Liquid invasion damage into the rock matrix is one of the major factors that cause low productivity in tight gas reservoirs [21]. In the absence of external cake protection, filtrate invasion into the tight formations is huge due to the tremendous amount of capillary pressure suction that potentially imbibes and holds the invaded water in the porous media [22]. The liquid phase trapping eventually reduces the near wellbore permeability as shown in Fig. 1.3 as a result of the temporary or permanent trapping of liquid inside the porous media [10]. In addition, liquid invasion into the tight formations normally continues for a noticeably long period of time as a result of weak mud cake development on wellbore wall. The weak mud cake and strong capillary pressure suction may amplify the water invasion profile and deteriorate severity of the phase trap damage to the tight formation. The greater the difference between initial water saturation and critical water saturation in tight formations, the more significant is the potential damage to gas permeability [23].

A typical tight gas reservoir may produce mainly dry gas, and contain very low amounts of heavy components. Therefore, condensate banking may not be a major issue in the tight gas wells, and the well can be allowed to flow at low flowing bottom-hole pressure for higher gas production rate. In addition, producing gas at low flowing bottom-hole pressure may reduce the water phase trapping damage near the wellbore, since at lower pressure, water content of the gas phase increases and water phase may partially be vaporized into the flowing gas phase in the reduced pressure zone around the wellbore [24]. However in the cases that are not

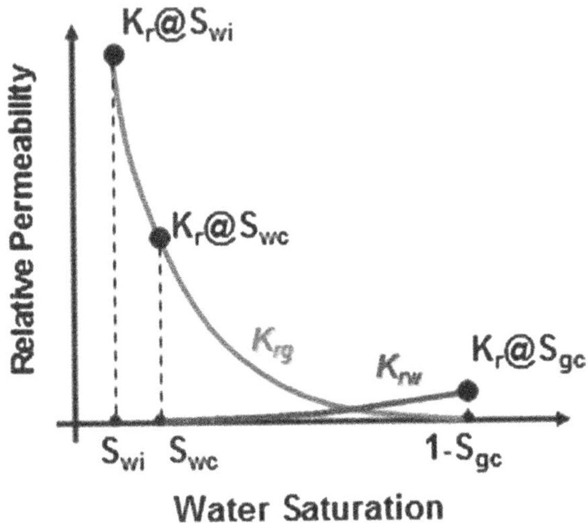

Fig. 1.3 Reduced effective permeability due to phase trapping damage [23]

truly dry gas situation, producing with the large pressure drawdown may cause condensate banking in the reservoir near the wellbore, if the flowing bottom-hole pressure drops below the dew point pressure of the gas phase [25].

Oil based fluids may be considered in some situations for low permeability gas reservoirs. In the case of oil-based drilling fluid invasion, there is no external water being introduced into the formation and the fluid saturations and wettability remain unchanged. However invasion of the oil filtrate into the tight formations may result in introduction of an immiscible liquid hydrocarbon around wellbore, causing entrapment of an additional third phase in the porous media. In the case of oil-based fluids invasion into water wet gas reservoirs, the invaded oil may tend to be trapped in the central portion of the pore space, rather than adhering tightly to the matrix walls as the wetting phase. Although this central pore space occlusion can cause substantial reductions in permeability, in some cases, the damaging effect in overall is less than the case where water based system is used in the same circumstances. Some types of oil may also dissolve in the gas and clean up after some time. The relative permeability curves in Fig. 1.4 show the reduced effective permeability due to the water invasion into the formation, compared with damage to permeability caused by oil invasion [23, 26].

1.4 Hydraulic Fracturing in Tight Gas Reservoirs

Hydraulic fracturing is performed to bypass the damaged zone and create larger contact area between the wellbore and the permeable conduits in the reservoir. The importance of hydraulic fracturing in tight gas sandstone reservoirs is well documented and understanding the hydraulic fracture parameters is essential for evaluation of the well production performance [27]. Massive hydraulic fractures in

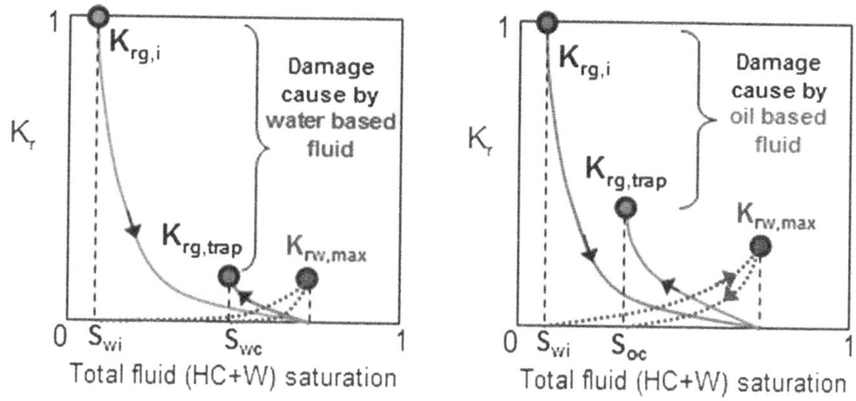

Fig. 1.4 Damage caused by water and oil phase trapping [23]

particular, can enhance the effective permeability around wellbore and may connect the wellbore to the adjacent sand lenses that are not penetrated by the well [28].

Propagation and direction of hydraulic fractures in tight formations are mainly controlled by in situ stresses as shown in Fig. 1.5. Where there is high contrast between minimum and maximum horizontal stresses, the stimulation creates a narrow or linear fracture fairway, and where the stress contrast is low, wide or complex fracture geometry are created during the treatment [29].

A common practice in unconventional gas reservoirs is to drill a horizontal well, consider short perforation intervals, and then hydraulically fracture the formation in multi-stages to create a treated zone around each hydraulic fracture [30]. Considering a horizontal well in a normal faulting stress regime, the hydraulic fracture might be different as illustrated in Fig. 1.6. If the horizontal well is drilled in the direction of maximum horizontal stress, the longitudinal hydraulic fractures are likely to be initiated along the wellbore, and if the horizontal well is drilled in the direction of minimum horizontal stress, then the transverse hydraulic fractures are initiated perpendicular to the wellbore axis [31].

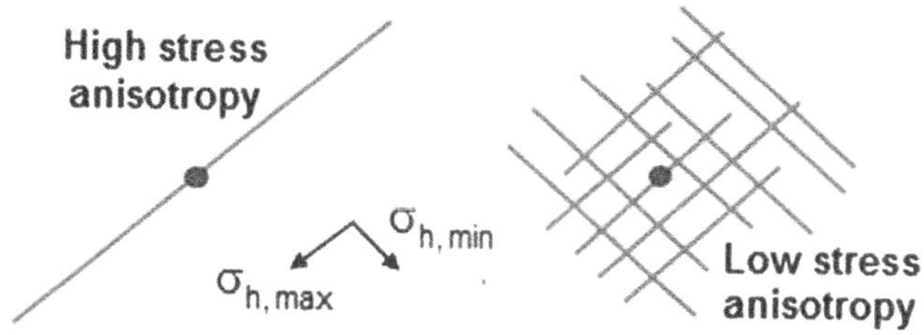

Fig. 1.5 Effect of stress anisotropy on propagation of hydraulic fractures

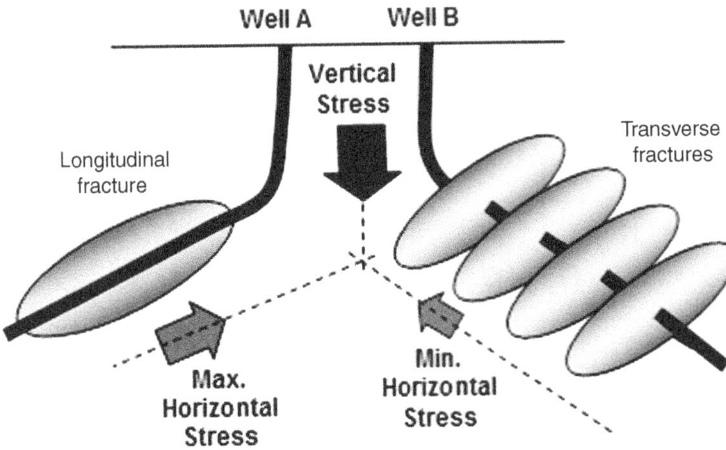

Fig. 1.6 Longitudinal and transverse hydraulic fractures in tight gas reservoirs

1.5 Damage Due to Hydraulic Fracturing

Hydraulic fracturing in some cases may not improve well productivity in tight gas reservoirs, or the productivity may increase only temporarily. During the stimulation and fracturing in tight gas reservoirs, fracturing liquids invade the reservoir and may create a bank of fracturing agent around the hydraulic fracture wings, which can develop negative effects on the long term production performance of the well [32]. Low productivity of a hydraulically fractured well might be due to the existence of damage mechanisms associated with liquid phase trapping in rock pores next to the fractures [8]. The tight formations with sub-normal initial water saturation are significantly more sensitive to damage caused by water phase trapping, and therefore water blocking may plague the success of hydraulic fracturing in low permeability gas reservoirs with this characteristic. The injected fluid during hydraulic fracturing should be compatible with formation to avoid clay swelling [9]. In the case of naturally fractured reservoirs, the fracturing fluids may transport the damaging solids through the natural fractures into deeper parts of the reservoir and further reduce the well productivity [33].

Use of polymer gels with hydraulic fracturing fluid may control the invasion of fracturing liquid and fluid loss into the tight formation, as the polymer is deposited on fracture faces, and cause very short distance penetration of the unbroken polymer gel into the reservoir rock. However this can make the fracturing fluid to be highly viscous, which may result in plugging as well as damaging of the hydraulic fractures face, dramatically lessen the effective length and width of the hydraulic fractures, and restrict the return of fluids during clean-up and gas production period [34]. The damage inside the fractures may also be due to proppant crushing, embedding, or fracture plugging with chemicals and polymers. The polymer may become a highly concentrated gel, and if it is left in the fractures, the gel damage can be the reason for ineffective clean-up and short effective fracture length. The polymer plugging can be reduced if suitable breakers are used, but this breaker must be able to be activated deep within the fractures [35].

In tight gas reservoirs that are sensitive to water invasion damage, hydraulic fracturing may fail to produce gas at commercial rates as it causes excessive liquid leak off into the tight formation. The preferred option in tight gas reservoirs might be horizontal well drilling in underbalanced conditions [36].

In hydraulic fracturing, additional production difficulties may also be experienced on the downstream side of the formation interface. These problems include proppant back-production that causes erosion of surface facilities [37]. Also in the case of significant liquid leak-off and fluid loss into the tight formation during fracturing, the post-fracturing gas production and the well productivity may be affected by loading of the fracturing liquid in wellbore that cannot be lifted to surface by the natural gas flow [38].

1.6 Thermal fracturing

During hydraulic stimulation using conventional fracturing fluid in tight gas reservoirs, water based fracturing fluid may eventually get trapped as liquid phase in rock pores next to the fractures due to very low permeability in tight formation, which can significantly damage near wellbore region in the form water blocking. Hence, the goal in tight formation stimulation should be to find a treatment which does not interact with the formation to avoid clay swelling, and minimize capillary effects.

One of the promising treatments is thermal fracturing by injection of a cold non-damaging liquid such as liquid form CO_2 into the tight gas reservoirs. The injection of cold liquid leads to cooling and thermal shrinkage in the formation around the injection well, which intern leads to occurrence of cracks (i.e. initiate the fracturing of formations). As the liquid CO_2 enters the formation, it gradually changes the state from liquid to gaseous, and therefore it does not cause any damage due to relative permeability and capillary pressure.

Reservoir rock can be stimulated by thermally induced stresses and exposing the hot formation rock to a cold fluid such as CO_2 in the liquid form. The sudden significant temperature change in the reservoir rock can break down the formation, create fractures, and result in improved well productivity. The thermal fracturing can be considered as an alternative stimulation plan, where hydraulic fracturing is not feasible. This process does not require high pump pressure as required in the hydraulic fracturing, and it is environmentally friendly.

Injection of cold CO_2 below the critical temperature as liquid, it can break down the brittle tight sand formation by thermally induced stresses. Injection of liquid CO_2 as a fracturing fluid in tight reservoirs offers a viable method of stimulation, and it can lead to creating several fractures due to the thermal stresses which may increase the effective permeability around the tight gas well. In addition, shrinkage resulting in cracking moves forward as the cooling front propagates far into the reservoir regions. Some field data where a cold liquid was injected into deep hot reservoirs, they indicated a sharp increase in injectivity after a certain time, almost as if the formation were fractured, while injecting pressure was below the formation breakdown pressure [39].

This effect can be considered as an alternative fracture stimulation treatments, especially where the conventional hydraulic fracture treatment job are found to be ineffective in the development of unconventional reservoirs. The amount of CO_2 required to achieve an optimum fracturing should be determined.

1.7 Mitigating Damage in Tight Gas Reservoirs

The damage mechanisms in tight gas reservoirs are controlled by pore system geometry, interfacial tension between the invading trapped fluid and the produced (or injected) reservoir fluid, capillary pressure, relative permeability, wettability, fluid saturation levels, depth of invading fluid penetration, reservoir temperature, reservoir pressure and well bottom-hole flowing pressure. With most of the phase trapping problems, prevention is generally more effective than remediation from an economic perspective. Removing damage is more common in the industry although it may be more problematic and certainly more costly [22, 23].

The damage due to liquid invasion and clay swelling can be minimized by properly choice of the base fluid for drilling and fracturing treatments, and reducing overbalance pressure during drilling and completion. Improving drilling or fracturing fluid rheology and filter cake building ability, which can provide an effective cake that is later removable can help control the damaged zone depth and reduce the damage due to liquid invasion. Reducing interfacial tension (IFT) between the trapped injected fluid and the reservoir fluid using IFT reducing agents such as methanol and liquid phase carbon dioxide can help more efficient recovery of the trapped phase from the invaded zone. Adding methanol in the fracturing fluid can reduce the water block as it helps faster clean-up and drying of water from the invaded zone [40, 41].

Using hydrocarbon-based drilling fluid in designing a drilling fluid can result in minimal phase trap potential, as it can avoid clay swelling. In addition, interfacial tension that directly affects capillary pressure and retention, it is significantly less between Diesel–Gas compared with Brine-Gas, and there using the oil-based drilling or fracturing fluid can reduce phase trapping damage. In a tight formation already damaged by water invasion, down-hole heating can be used to vaporize water in the invaded zone and remove the aqueous phase traps as well as thermally decomposing potentially reactive swelling clays. The water phase trapping may also be removed by injection of the dehydrated dry gas into the formation to initiate conduits of higher gas permeability through the damaged zone. Thermal fracturing using cold liquid CO_2 is another method of improving well productivity without causing damage to formation in tight gas reservoirs [23, 42].

1.8 Summary

Based on this literature review, it is evident that there are many factors that can influence the production performance of tight gas reservoirs. Understanding the effects can be paramount for successful development and exploitation of tight gas reservoir. This chapter reviewed the different factors that control damage mechanism and well productivity, and discussed the optimum strategies for developing the tight gas reservoirs.

References

1. Law BE and Curtis JB (2002) Introduction to unconventional petroleum systems. AAPG Bulletin 86:1851–1852
2. Holditch SA (2006) Tight gas sands. Paper presented at the SPE 103356, Texas A and M University. J Petrol Technol
3. Meeks MH, Susewind KD, Templeman TL (2006) Maximizing gas recovery from tight gas reservoirs in Trawick field. Paper presented at the SPE 101221, international petroleum exhibition and conference, Abu Dhabi
4. Fairhurst DL, Indriati S, Reynolds BW (2007) Advanced technology completion strategies for marginal tight gas sand reservoirs: a production optimization case study in South Texas. Paper presented at the SPE 109863, SPE annual technical conference and exhibition, California, USA
5. Campbell I (2009) An overview of tight gas resources in Australia, PESA News 95–100
6. Gonfalini M (2005) Formation evaluation challenges in unconventional tight hydrocarbon reservoirs. Paper presented at the 2005 SPE Italian section, Italy
7. Teufel LW, Chen HY, Engler TW, Hart B, (2004) Optimization of infill drilling in naturally-fractured tight-gas Reservoirs. Final report, by New Mexico Institute of Mining and Technology, submitted to U.S. Department of Energy and Industry Cooperative
8. Mahadevan J, Sharma M, Yortsos YC (2007) Capillary wicking in gas wells. Paper presented at the SPE 103229. SPE J 12(4):429–437
9. Bennion DB, Brent F (2005) Formation damage issues impacting the productivity of low permeability, low initial water saturation gas producing formations. J Energy Res Technol 127:240246
10. Bennion DB, Thomas FB (1996) Low permeability gas reservoirs: problems, opportunities and solutions for drilling, completion, stimulation and production. Paper presented at the SPE 35577, SPE gas technology conference, Calgary
11. Kantanong N, Bahrami H, Rezaee R, Hossain MM (2012) Effect of sand lens size and hydraulic fractures orientation on tight gas reservoirs ultimate recovery. Paper presented at the SPE 151037, middle east unconventional gas conference, Abu Dhabi
12. Bahrami H, Rezaee R, Hossain MM, Murickan G, Basharudin N, Alizadeh N, Fathi A (2012) Effect of sand lens size and hydraulic fractures parameters on gas in place estimation using 'P/Z vs Gp method in tight gas reservoirs. Paper presented at the SPE 151038, SPE/EAGE European unconventional resources conference and exhibition, Vienna, Austria
13. Jaeger JC, Cook NGW, Zimmerman RW (2007) Fundamentals of rock mechanics 4th edn, Blackwell Chap 12, p 370
14. Abass H, Ortiz I (2007) Understanding stress dependant permeability of matrix, natural fractures, and hydraulic fractures in carbonate formations. Paper presented at the SPE 110973, SPE technical symposium, Saudi Arabia

15. Teufel LW, Sandia N (1993) Control of fractured reservoir permeability by spatial temporal variations in stress magnitude and orientation. Paper presented at the SPE 26437, SPE annual technical conference, Texas
16. Dusseault MB (1993) Stress changes in thermal operations, SPE 25809, SPE International Thermal Operations Symposium, Bakersfield, California
17. Behrmann L (2000) Perforating practices that optimize productivity, oilfield review. Spring
18. Holditch SA (1979) Factors affecting water blocking and gas flow from hydraulically fractured gas wells. JPT J 31(12):1515–1524
19. Sharif A (2007) Tight-gas resources in the Western Australia. PWA
20. Araujo EM, Fontoura SA, Pastor JS (2005) A methodology for drilling through shales in environments with narrow mud-weight, SPE 94769, 2005 SPE Latin American and Caribbean Petroleum Engineering Conference
21. You L, Kang Y (2009) Integrated evaluation of water phase trapping damage potential in tight gas reservoirs. Paper presented at the SPE 122034, European formation damage conference, Scheveningen, The Netherlands
22. Ding Y, Herzhaft B, Renard G (2006) Near-wellbore formation damage effects on well performance: a comparison between underbalanced and overbalanced drilling. SPE Prod Oper 21:1
23. Bennion DB, Thomas FB, Schlumeister B, Romanova UG (2006) Water and oil base fluid retention in low permeability porous media—an update. Paper presented at the canadian international petroleum conference, Calgary
24. Lokken VT, Bersas A, Christensen KO, Nygaard CF, Solbraa E (2008) Water content of high pressure natural gas. Paper presented at the international gas union research conference, Paris
25. Ravari RR, Wattenbarger RA, Ibrahim M (2005) Gas condensate damage in hydraulically fractured wells. Paper presented at the SPE 93248, Asia pacific oil and gas conference and exhibition, Indonesia
26. Chi S, Torres C, Wu J, Alpak OF (2004) Assessment of mud-filtrate invasion effects on borehole acoustic logs and radial profiling of formation elastic parameters. Paper presented at the SPE 90159, SPE annual technical conference, Houston
27. Wang Y (2008) Simulation of fracture fluid clean-up and its effect on long term recovery in tight gas reservoirs, Ph.D. thesis report, Texas A&M University
28. Cipola C, Mack M (2010) Reducing exploration and appraisal risk in low-permeability reservoirs using microseismic fracture mapping. Paper presented at the SPE 137437, Canadian unconventional resources and international petroleum conference, Calgary
29. Fan L, Thompson JW, Robinson JR (2010) Understanding gas production mechanism and effectiveness of well stimulation in the Haynesville shale through reservoir simulation. Paper presented at the SPE 136696, Canadian unconventional resources and international petroleum conference, Calgary, Alberta, Canada
30. Bagherian B, Sarmadivaleh M, Ghalambor A, Nabipour A, Rasouli V, Mahmoudi M (2010) Optimization of multiple-fractured horizontal tight gas well. Paper presented at the SPE 127899, symposium and exhibition on formation damage control, Lafayette, Louisiana
31. Hossain MMM, Rahman MK (2008) Numerical simulation of complex fracture growth during tight reservoir stimulation by hydraulic fracturing. J Petrol Sci Eng 60:86–104
32. Wang JY, Holditch SA and Duane A (2010) Modeling fracture-fluid cleanup in tight-gas wells, SPE 119624, SPE J 15(3)
33. Rodgerson JL (2000). Impact of natural fractures in hydraulic fracturing of tight gas sands. Paper presented at the SPE 59540, Permian basin oil and gas recovery conference, Midland, Texas
34. Raible CJ, Gall BL (1985) NIPER, laboratory formation damage studies of western tight gas sands. Paper presented at the SPE 13903
35. Wang Y, Holditch SA, (2008) Simulation of gel damage on fracture fluid cleanup and long-term recovery in tight gas reservoirs. Paper presented at the SPE 117444, SPE eastern regional/AAPG eastern section joint meeting, Pennsylvania

36. Veeken CAM, Velzen JFG, Beukel JVD, (2007) Underbalanced drilling and completion of sand prone tight gas reservoirs in North Sea. Paper presented at the SPE 107673, European formation damage conference, The Netherlands
37. Abass H (2009) Optimizing proppant conductivity and number of hydraulic fractures in tight gas sand wells. Paper presented at the SPE 126159, SPE technical symposium, Saudi Arabia
38. Salim P, Li J, (2009) Simulation of liquid unloading from a gas well with coiled tubing using a transient software. Paper presented at the SPE 124195, SPE annual technical conference and exhibition, New Orleans
39. Mueller M, Amro M, Karl Haefner F, Hossain MM (2012) Stimulation of tight gas reservoir using coupled hydraulic and CO_2 cold-frac technology, SPE 160365, Asia Pacific oil and gas conference and exhibition, Perth, Australia
40. Bazin B, Peysson Y, Lamy F, Martin F, Aubry E, Chapuis C (2009) Insitu water blocking measurement and interpretation related to fracturing operations in tight gas reservoirs. Paper presented at the SPE formation damage control symposium, The Netherlands
41. Motealleh S, Bryant SL (2009) Permeability reduction by small water saturation in tight gas sandstones. SPE J 14(2):252–258
42. Jamaluddin AKM, Mehta SA, Moore RG (1998) Downhole heating device to remediate near-wellbore formation damage related to clay swelling and fluid blocking, Document ID: 98–73. Paper presented at the SPE Annual Technical Meeting, Calgary, Alberta

Chapter 2
Tight Gas Reservoirs Characterisation for Dynamic Parameters

Tight gas reservoirs might be very different in term of reservoir characteristics, and it is challenging to adequately determine the reservoir dynamics parameters such as the effective permeability of matrix and natural fractures, relative permeability, skin factor, hydraulic fractures size and conductivity and fluid gradients in the reservoir. Similar to the conventional gas reservoirs, the reservoir characterization tools such as well testing, logging, core analysis and formation testing are commonly used and run in tight gas reservoirs. However due to the tight formations complexity, heterogeneity and very low permeability, use of the acquired data to obtain meaningful results may not be well understood in term of determining the well and reservoir parameters and predicting the well production performance [13, 14]. This chapter presents the methods for determination of the effective permeability of non-fractured, hydraulically fractured and naturally fractured tight formations.

2.1 Welltest Analysis in Tight Gas Reservoirs

The conventional method of pressure build-up test data analysis is to plot the transient pressure (P) and its derivative (P':$-d[\Delta P]/d[Log((t_p + \Delta t)/\Delta t)]$) versus time on Log–Log scales to identify the radial flow regime and determine the reservoir permeability (KAPPA engineering 2011). Diagnosis of the radial flow regime is critical in quantitative welltest interpretation, since reliable values for reservoir permeability and skin factor can be estimated when radial flow regime is established in the reservoir [3, 9].

A pressure transient test breaks into several flow regimes, each seeing deeper in the reservoir than the last [9]. Depending on well completion type, completion configuration, reservoir geological and geometric attributes, different flow regime might be revealed on diagnostic plots of pressure transient data analysis in vertical, horizontal or multi-fractured wells [5]. In pressure transient tests, the pressure derivative curve can indicate the different flow regimes: the slope of +1 shows wellbore storage effect, the slopes of −0.5, +0.5, +0.25 and +0.36 indicate

N. Bahrami, *Evaluating Factors Controlling Damage and Productivity in Tight Gas Reservoirs*, Springer Theses, DOI: 10.1007/978-3-319-02481-3_2,

spherical, linear, bi-linear and elliptical flow regimes respectively, and the slope of zero indicates radial flow regime.

In multi-fractured tight gas wells, the main reservoir flow regimes are the early time linear flow regime (slope of +1/2 on pressure derivative) perpendicular to the hydraulic fractures inside Stimulated Reservoir Volume (SRV), followed by the early-time elliptical flow regime towards the drainage area of the linear flow (slope of +1/3 on pressure derivative), and then the early-time boundary dominated flow when drainage area around the hydraulic fractures is depleted. After depletion of SRV, gas flow is provided by the untreated rock surrounding SRV, which acts as boundary. The early time boundary dominated flow effect is then followed by the late-time linear flow regime and then late-time elliptical flow regime inside the untreated reservoir rock towards the SRV around the multi-fractured well. Finally at very late-time when pressure disturbance propagates deep enough into the reservoir, a pseudo radial flow regime is established with slope of zero on pressure derivative [4]. The linear and elliptical flow regimes may be the dominant flow regimes in tight gas wells with significantly long time duration and the radial flow regime may not be reached, due to the very low reservoir permeability, heterogeneity, hydraulic fractures, natural fractures, permeability anisotropy, and reservoir geometry [18, 2].

The early portion of welltest data during pressure build-up tests is normally affected by wellbore storage and skin factor. In tight gas reservoirs, the low permeability slows down the reservoir response to the pressure disturbance during transient testing, which causes the wellbore storage effect to be significantly long [12]. As a result, tight gas reservoirs typically require a relatively long pressure build-up testing time to reach the late time pseudo radial flow regime, which is often not practical. In addition, the need for hydraulic fracturing to obtain commercial flow rates in tight gas reservoirs adds to the complexity of the problem and makes analysis of the pressure transient data more difficult. Therefore, welltest analysis using the conventional techniques may fail to provide reliable results.

2.2 Permeability Estimation in Hydraulically Fractured Tight Gas Wells

The main challenge with welltest analysis of the tight gas wells is that the testing time cannot be long enough to reach radial flow regime. However for a multi-stage fractured horizontal tight gas well, if both of the linear flow and elliptical flow regimes are detected on the Log–Log diagnostic plot, they can be used in welltest analysis to estimate reservoir characteristics.

Welltest analysis of hydraulically fractured tight gas wells requires plotting of the pressure transient data and the derivatives on the Log–Log diagnostic plot: $d(p)/d(t^{1/2})$ is defined as the linear flow derivative, $d(p)/d(t^{1/3})$ is defined as the elliptical flow derivative, and $d(p)/d(\ln[t])$ is defined as the radial flow derivative.

Then the pressure derivative values for linear flow, elliptical flow and radial flow regimes can be determined as shown in Fig. 2.1: m_{LF} from zero slope line on linear derivative, m_{Ell} from zero slope line on elliptical derivative and m_{RF} from zero slope line on radial derivative [15, 10].

The linear flow and elliptical flow regimes are both controlled by K and X_f. The solution of the diffusivity equation for infinite acting elliptical flow has been proposed for non-fractured horizontal oil and gas wells [15]. Modifying and re-deriving the elliptical flow equation for a hydraulically fractured horizontal gas well, and integrating that with the linear flow equation can provide the hydraulic fracture size and reservoir permeability as follows:

$$X_f = \frac{573}{K^{0.89}\,(\phi\mu C_t)^{0.5}}\left(\frac{qT}{h * n * m_{Ell}}\right)^{1.39} \tag{2.1}$$

$$K = \left(\frac{40.99qT}{h * n * X_f * m_{LF}}\right)^{2}\frac{1}{\varphi\mu C_t} \tag{2.2}$$

Solving the two non-linear equations simultaneously can provide the two unknowns K and X_f. In the above equations, P is pressure (psia), t is time (hrs), q is gas flow rate (MSCFD), B is formation volume factor, h is reservoir thickness (ft), μ is viscosity (cp), φ is porosity (fraction), C_t is total compressibility, T is reservoir temperature (R), K is reservoir permeability (md), and n is number of fractures.

On the pressure build-up diagnostic plots for a multi-stage fractured horizontal tight gas well, if only the linear flow regime is detected and the testing time is not long enough to reach elliptical and radial flow regimes, then alternative methods such as using the second derivative of transient pressure should be used to proceed with welltest analysis and predict a theoretical radial flow regime. The theoretical radial flow should be predicted to estimate m_{RF} value, assuming an infinite

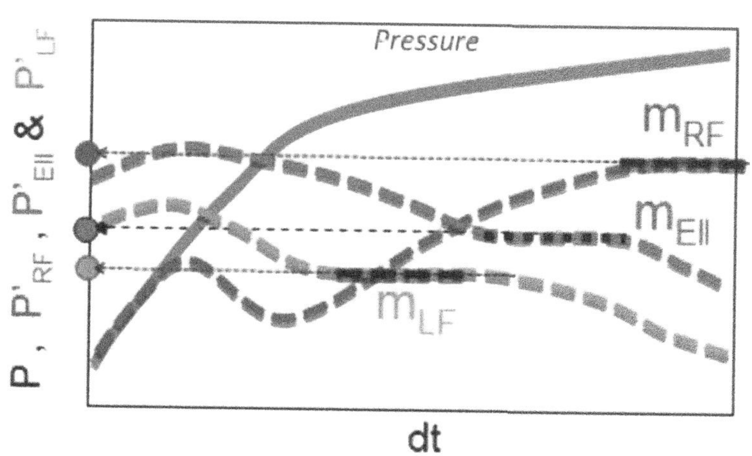

Fig. 2.1 Welltest analysis using the pressure derivatives

stimulated reservoir volume [4]. Using value of the linear flow derivative, m_{LF}, and the value of m_{RF} from the theoretical radial flow prediction, the reservoir permeability and hydraulic fracture half-length size can be estimated as follows:

$$K = \frac{1637 * Q_g * T}{2.3 * m_{RF} * h * n} \tag{2.3}$$

$$X_f = \frac{40{\cdot}99\, Q_g T}{h * m_{LF} * n} \sqrt{\frac{1}{K\varphi\mu C_t}} \tag{2.4}$$

Permeability estimation using the method may have significant uncertainties, due to some assumptions in predicting the theoretical radial flow regime.

2.3 Estimating Permeability of the Natural Fractures

Natural fractures may contribute the most to total gas production from tight gas reservoirs, and therefore identification of their characteristics is essential for well production performance evaluation.

The basic dynamic characteristics of the natural fractures are fracture storativity and interporosity flow coefficient, which can be estimated from welltest analysis. Then using the parameters, natural fractures permeability can be estimated as follows [22]:

$$K_f = \delta \frac{K_m}{\lambda} r_w^2 \tag{2.5}$$

$$\delta = 4(1/a_X^2 + 1/a_Y^2 + 1/a_Z^2) \tag{2.6}$$

where K_m is matrix permeability, K_f is fracture permeability, r_w is wellbore radius, δ is shape factor, and λ is interporosity flow coefficient. a_x, a_y and a_z are matrix block size respectively in x, y and z directions. In the case of Kazemi model ($a_\chi \gg a_z$ and $a_y \gg a_z$), the shape factor, δ, is considered to be $4/a_z^2$. The shape factor can be estimated from image log fracture spacing (a_z); matrix permeability can be estimated from core analysis; and the interporosity flow coefficient can be estimated from welltest analysis if dual-porosity, dual-permeability response is clearly observed on pressure build-up diagnostic plots [7].

However in tight gas reservoirs, due to the long wellbore storage effect and also the tightness and heterogeneity of the reservoir rock, pressure build-up diagnostic plots may not be able to show the dual porosity dual permeability response. Hence, estimating the interporosity flow coefficient and fracture permeability from such welltest data might not be feasible, and the conventional approaches might fail to characterize the fracture parameters in tight gas reservoirs.

Permeability of natural fractures for tight gas reservoirs can be estimated based on Kazemi model that assumes parallel layers of matrix and fracture in a uniform fracture network model [23] and averaging the reservoir permeability based on thickness of matrix and fracture layers [7, 9]:

$$K \cdot h_{Average} = \sum_{Matrix}^{m=1\ldots n} (K_m * a) + \sum_{fracture}^{f=1\ldots n} (K_f * b) \tag{2.7}$$

$$h = (n * \mathrm{a}) + (n * \mathrm{b}) \tag{2.8}$$

where K_f is permeability of a natural fracture, b is average fracture aperture, a is average matrix block thickness, K is welltest permeability, K_m is average permeability of the matrix blocks, h is reservoir thickness, n is number of fractures intersecting the wellbore across the reservoir, φf is fracture porosity (fraction), n * a is cumulative matrix block thickness, and n * b is cumulative fracture aperture. By combining Eqs. 2.6 and 2.7 using the assumption of $a \gg b$, $K_f \gg K_{welltest}$ and $K_f \gg K_m$ for tight gas reservoirs, following simplified equation can be written:

$$K_f = K_{welltest} * \frac{a}{b} \tag{2.9}$$

Since Eq. 2.8 is based on simplified models and assumptions, using some correction factors might provide more realistic relationship between fracture dynamic parameters. Considering the correction factors, average permeability of natural fractures ($K_{f,average}$) can be expressed in the following generalized form:

$$K_{f,\,average} = C_1 * K_{welltest} * \left(\frac{a_f}{b_f}\right)^{C_2} \tag{2.10}$$

Where $K_{welltest}$ is welltest permeability, b_f is average fracture aperture, a_f is average fracture spacing, and C_1 and C_2 are the correction factors. For a tight gas reservoir, average permeability can be estimated from welltest analysis, fracture spacing and fracture aperture can be approximated from image log processing, and the constants C_1 and C_2 can be determined from sensitivity analysis using reservoir simulation models, or as matching parameter during field history matching.

It should also be noted that where there is significant in situ stress anisotropy, permeability of the natural fractures would be different in different directions [6]. The natural fractures that are aligned with maximum horizontal stress (perpendicular to the minimum stress direction) may have larger aperture and therefore greater permeability, compared with the natural fractures perpendicular to the maximum stress direction. As function of in situ stresses, the maximum permeability and horizontal permeability can be estimated as follows [5]:

$$K_{f,\max} = K_{f,average} * \left(\frac{\sigma_{\min}}{\sigma_{\max}}\right)^{\alpha/2} \tag{2.11}$$

$$K_{f,\min} = K_{f,\,average} * \left(\frac{\sigma_{\max}}{\sigma_{\min}}\right)^{\alpha/2} \tag{2.12}$$

where $K_{f,average}$ is average permeability of natural fractures, $K_{f,min}$ is permeability of the natural fractures that are perpendicular to the maximum stress direction, $K_{f,max}$ is permeability of the natural fractures that are perpendicular to the minimum stress direction, σ_{min} is minimum stress, σ_{max} is maximum stress, and α is a constant number that can be estimated from core analysis experiments and plotting permeability versus normal stress. For a typical tight gas reservoir in Western Australia, the value of α was estimated as -1.28 [5].

2.4 Relative Permeability Curves in Tight Gas Reservoirs

The major damage mechanisms in tight gas reservoirs such as phase trapping are found to be associated with relative permeability and capillary pressure curves. The damaging effects are reflected on gas and water relative permeability curves [8].

The relative permeability data for tight gas sands are extremely difficult to obtain by the conventional steady state flow analysis technique as it requires impractically very long stabilization time and flow rates are usually small [16]. However, an unsteady state technique can be applied in a core flooding experiment, in which the tight core samples are fully saturated with water (initial water saturation of 100 % for primary drainage), and then gas is flooded at constant volumetric flow rate to reach irreducible water saturation. During the core flooding experiment, the pressure differential across the core sample and volume of the produced water are recorded. Then the experimental core flood data can be input into a commercial core flooding data analysis software, in order to generate relative permeability curves by matching the core flood data for brine production and pressure differential [19].

For oil-gas system, relative permeability curve can be generated similarly. However if determination of gas-oil relative permeability may not be possible due to some limitations in the laboratory facilities when oil is used as liquid phase, then typical published oil-gas relative permeability data have to be considered in the reservoir simulation studies related to oil-gas system [17].

2.5 Pressure Measurement in Tight Gas Reservoirs

In order to measure the pressure of a formation, pressure gradient, and gas water contact, formation testing is used. To measure pressure of reservoir at each depth, the tool inserts to a probe into the borehole wall to perform a mini pressure drawdown and build-up by withdrawing a small amount of formation fluid, and then waiting for the pressure to build up to the formation pore pressure at that depth. Formation testers measure the pressure of the continuous phase in the invaded region, which is the pressure of the drilling fluid filtrate. Using the pressure measurements at different depths, gradient of pressure in the reservoir is determined, which can indicate reservoir fluid type and water-hydrocarbon contact [20].

In tight gas reservoirs, formation testing is challenging due to tightness of the reservoir rock, weak mud cake across the wellbore, and presence of large wellbore breakouts across the tight sand intervals. Although using advanced formation testing tools may help improve reservoir characterization of tight gas reservoirs [21], formation testing results in tight formations may still have some uncertainties. In good permeability zones, formation tests are effective and normal. However in the case of low reservoir permeability, the mud cake is often ineffective in preventing filtrate invasion, thus causing the measured pressure to be affected by wellbore pressure that might be higher than the actual formation pressure (supercharging effect). In testing of a very tight formation, even a large pressure drawdown may result in no flow from the reservoir (dry test). Tight gas reservoirs are often associated with bad-hole conditions (large wellbore breakouts) causing lost seals around the tool packer and failure during testing of the formation [20]. The formation testing measurements may also be influenced by the effects of capillary pressure in the case of liquid invasion into a gas bearing zone. As a result, the measured pressure might be different to the true formation pressure [1, 11].

2.6 Summary

The tight gas reservoirs dynamic parameters such as relative permeability, reservoir average permeability, and natural fractures permeability are the key factors that control production performance of tight gas wells. This section presented a new method of welltest analysis for more reliable estimation of the average reservoir permeability and a new correlation for estimating the permeability of natural fractures in tight formations.

References

1. Andrews JT, Bahrami H, Rezaee R, Pourabed H, Mehmood S, Salemi H (2012) Effect of liquid invasion and capillary pressure on wireline formation tester measurements in tight gas reservoirs, SPE-154649, IADC/SPE Asia pacific drilling technology conference and exhibition held in Tianjin, China
2. Arevalo JA, Wattenbarger RA, Samaniego-Verduzco F, Pham TT (2001) Production analysis of long-term linear flow in tight gas reservoirs: case histories, SPE 71516, Annual Technical Conference and Exhibition, New Orleans, Louisiana

3. Badazhkov D (2008) Analysis of production data with elliptical flow regime in tight gas reservoirs, SPE 117023. SPE Russian Oil and gas technical conference and exhibition, Moscow
4. Bahrami H, Siavoshi J (2013) Interpretation of reservoir flow regimes and analysis of, SPE-164033. Middle East Unconventional Gas Conference and Exhibition held in Muscat, Muscat
5. Bahrami H, Mohemad Nour J, Alwerfaly Kh, Mutton G, Owais SA, Al-Waley A (2013) Whicher range field development planning, Study project in FDP course (MSc level, semester-3) in Curtin University, Perth
6. Bahrami H, Rezaee R, Kabir A, Siavoshi J, Jammazi R (2010) Using second derivative of transient pressure in welltest analysis of low permeability gas reservoirs, SPE 132475. Production and Operations Conference, Tunis
7. Bahrami H, Rezaee R, Hossain MM, Murickan G, Basharudin N, Alizadeh N, Fathi A (2012) Effect of sand lens size and hydraulic fractures parameters on gas in place estimation using 'P/Z vs Gp Method' in tight gas reservoirs, SPE 151038. SPE/EAGE European Unconventional Resources Conference and Exhibition held in Vienna, Austria
8. Bennion DB, Thomas FB, Schlumeister B, Romanova UG (2006) Water and oil base fluid retention in low permeability porous media—an update. Canadian International Petroleum Conference, Calgary
9. Bourdarot G (1998) Well testing interpretation methods, Editions Technip, Paris
10. Dynamic data analysis book, 2011, KAPPA Engineering
11. Elshahavi H, Fathy K, Heikal S (1999) Capillary pressure and rock wettability effects on wireline formation tester measurements, SPE 56712, SPE Annual technical conference and exhibition, Houston
12. Garcia JP, Pooladi Darvish M, Brunner F, Santo M, Mattar L (2006) Welltesting of tight gas reservoirs, SPE 100576, SPE Gas technology symposium, Calgary
13. Gonfalini M (2005) Formation evaluation challenges in unconventional tight hydrocarbon reservoirs, presented in 2005 SPE Italian section, Italy
14. Mahadik MK, Bahrami H, Hossain M, Mitchel PAT (2012) Production decline analysis and forecasting in tight-gas reservoirs APPEA Journal
15. Martinez, J. A., Escobar, F.H., Bonilla, L.F. (2012). Formulation of the elliptical flow for welltest interpretation in horizontal wells. ARPN J Eng Appl Sci vol 7(3), March 2012
16. Ning X, Holditch SA (1990) The measurement of gas relative permeability for low permeability cores. Annual SCA technical conference, Dallas, Texas, USA
17. Ravari RR, Wattenbarger RA, Ibrahim M (2005) Gas condensate damage in hydraulically fractured wells, SPE 93248, Asia Pacific Oil and gas conference and exhibition, Indonesia
18. Restrepo DP, Tiab D (2009) Multiple fractures transient response, SPE 121594. Latin American and Caribbean Petroleum Engineering Conference, Cartagena, Colombia
19. Saeedi Ali (2012) Experimental study of multi-phase flow in porous media during CO2 geo-sequestration processes, Springer Theses
20. Schlumberger Formation Testing (2005) Schlumberger reservoir log analysis training course handouts. Abu Dhabi
21. Schrooten RA (2007) A case study: using wireline pressure measurements to improve reservoir characterization in tight formation gas field, IPTC 11545. International petroleum technology conference, Duabi
22. Tiab D, Restrepo DP, Lgbokoyi A (2006) Fracture porosity of naturally fractured reservoirs, SPE 104056. First International Oil Conference and Exhibition in Mexico, Mexico
23. Van Golf-Racht TD (1982) Fundamentals of fractured reservoir engineering, 1st edn. Elsevier

Chapter 3
Tight Gas Reservoir Simulation

Analytical and numerical simulation studies are performed to have a qualitative understanding of damage mechanisms associated with production from non-fractured and hydraulically fractured tight gas reservoirs; and evaluate its potential impact on well productivity.

In building tight gas reservoirs simulation model, it is important to use a consistent set of field data in order to get meaningful simulation outputs. Based on a West Australian tight gas field data, the simulation models are built at reservoir scale and core scale. In this chapter, reservoir simulation studies for different types of well and tight reservoirs are presented.

3.1 Effect of Damage Mechanisms on Well Productivity

Reservoir simulation is used to understand how damage mechanisms are controlled by the well and reservoir parameters such as reservoir permeability, permeability of the damaged zone, radius of the damaged zone, drilling fluid type, capillary pressure and relative permeability curves.

In this section, the effects of different parameters on damage and skin factor are studied using the reservoir simulation models. To evaluate the damage effects, the term flow efficiency (FE) is used in some of the cases, which is defined as the ratio of the pressure drop across the model in the case of zero skin virgin homogeneous rock; to the pressure drop in the case of perforated and/or damaged rock (FE equals to 1 in the case that there is no damage introduced to a non-perforated model).

3.1.1 Damaged Zone Permeability and Radius

The simulation model is run for conventional and tight cores, with damaged zone permeability of K_d and damaged zone radius of r_d. The model results are shown in

N. Bahrami, *Evaluating Factors Controlling Damage and Productivity in Tight Gas Reservoirs*, Springer Theses, DOI: 10.1007/978-3-319-02481-3_3,

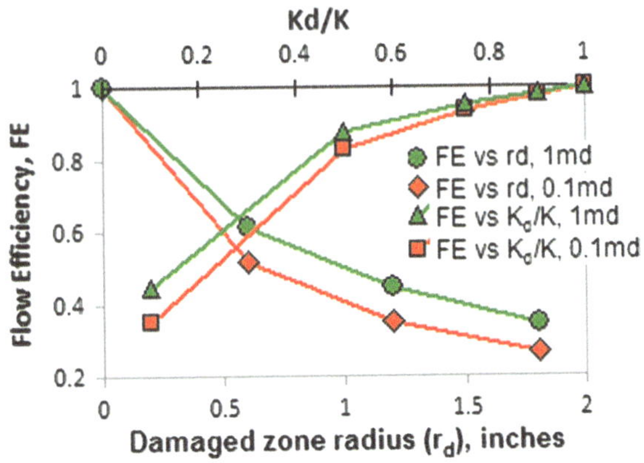

Fig. 3.1 Invaded zone parameters and the effect on flow efficiency

Fig. 3.1. According to the results, the effect of damaged zone permeability and damaged zone radius on flow efficiency is more significant in tight gas reservoirs compared with conventional cores, indicating the importance of damage control in tight gas reservoirs.

3.1.2 Phase Trapping Damage Caused by Water Invasion

The effect of water invasion in the reservoir model is evaluated by injecting water at the well location, followed by gas production. The water saturation in the reservoir model at initial conditions (top view) is shown in Fig. 3.2 ($S_{wi} = 0.6$).

First, water is injected at the well location, which increases water saturation around the wellbore. Water saturation at the end of the injection period is shown in

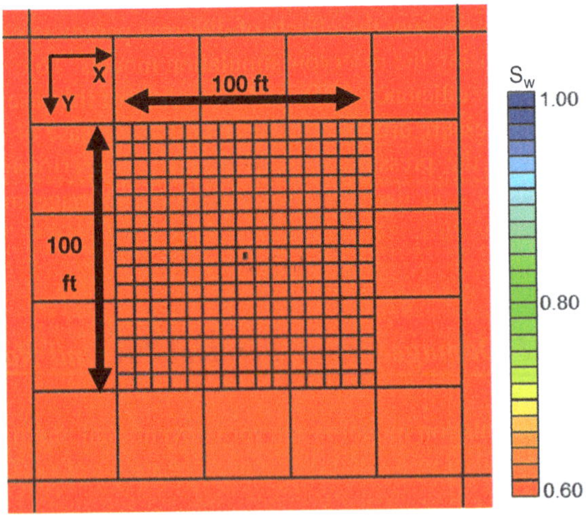

Fig. 3.2 Water saturation in the model before water invasion

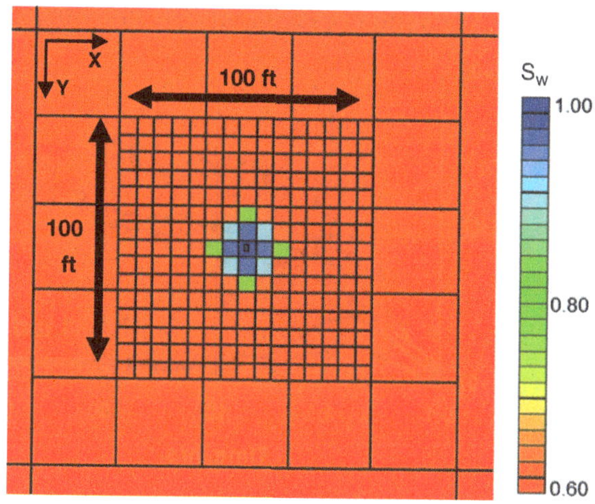

Fig. 3.3 Water saturation in the model at the end of water injection period

Fig. 3.3 (equivalent radius of water invaded zone: 9 ft). Afterwards, the model is put on gas production to clean-up the invading water, and reduce water saturation around the wellbore. Water saturation at the end of the gas production period is shown in Fig. 3.4 (equivalent radius of water invaded zone: 12 ft). The results indicate during the gas production phase, not only water from the near wellbore was not cleaned up by gas production, water invasion was continued into the reservoir due to the strong capillary pressure suction effects, and damaged zone radius (water invaded radius) increased with passage of time.

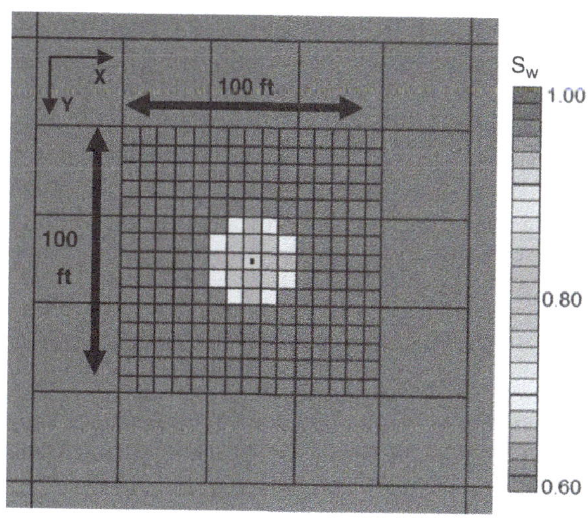

Fig. 3.4 Water saturation in the model at the end of gas production period

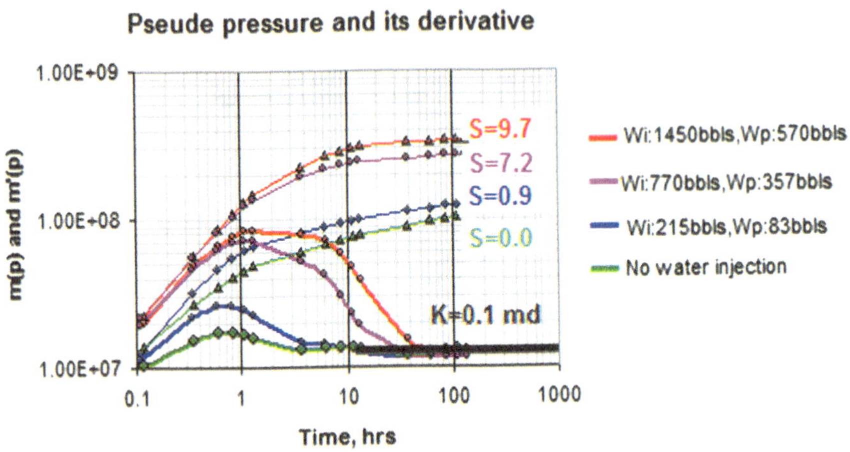

Fig. 3.5 Effect of phase trap damage on skin factor

3.1.3 Effect of Phase-Trap Damage on Skin Factor

The reservoir model is also run to understand the effect of phase trapping damage on skin factor for four different cases. Case A considers no leak-off of liquid into formation (no damage). Cases B, C and D, consider, respectively approximately 215, 770 and 1400 barrels of water leaks off into the formation. In each run, the water leak-off is followed by gas production during clean-up.

In each of the models after the liquid leak-off, the well is put on gas production followed by a pressure build-up test. The pressure transient data are generated to calculate the skin factor caused by phase trapping. The cumulative injected volume of water during leak-off (W_i) and the simulated results for cumulative produced water (W_p) during clean-up and gas production are integrated with welltest results as shown in Fig. 3.5. In the case of no liquid leak-off into the tight formation (case A), the water blocking skin is zero. In the case of significant water leak-off into the formation, skin is found to be positive. The results highlight the fact that phase trap related damage due to water leak-off into the tight gas reservoir causes positive skin factor, and significant reduction in gas production rate and gas recovery.

3.1.4 Overbalanced and Underbalanced Drilling

The model is run at core scale, to understand the effect of wellbore pressure on water invasion during overbalanced, balanced and underbalanced drilling. The model is run for the following cases.

- 500 psia overbalanced pressure resulted in 0.5″ liquid invasion into matrix

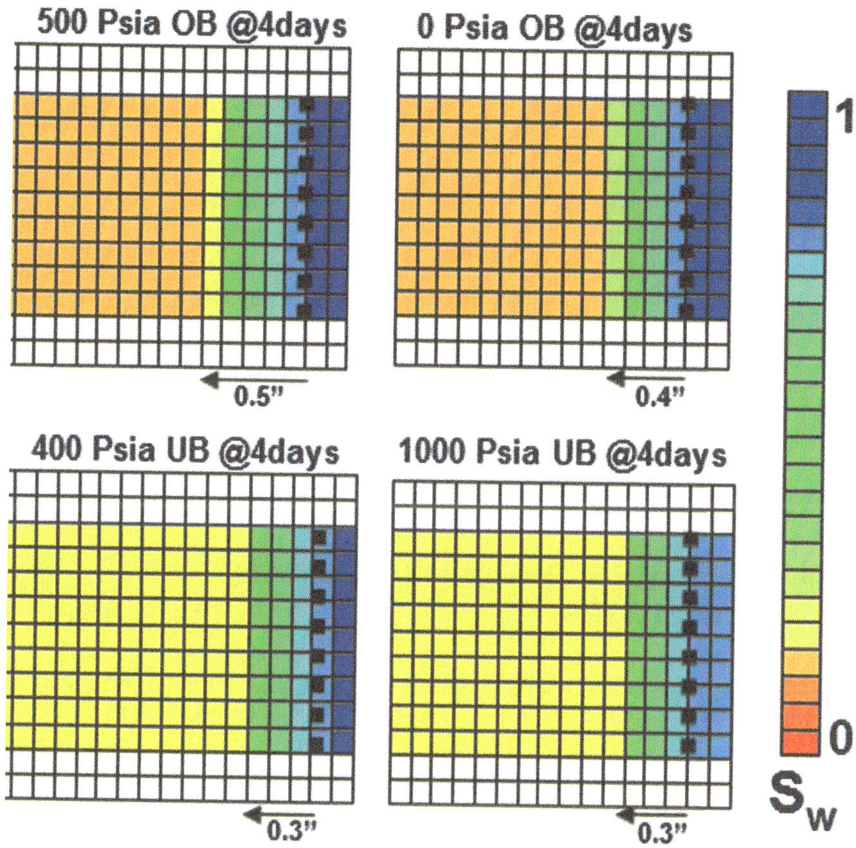

Fig. 3.6 Effect of wellbore pressure during drilling on phase trap damage

- Balanced pressure conditions resulted in 0.4″ liquid invasion into matrix
- 400 psia underbalanced resulted in 0.3″ liquid invasion into matrix
- 1000 psia underbalanced resulted in 0.3″ liquid invasion into matrix.

From this simulation results as shown in Fig. 3.6, it is obvious that the wellbore liquid invades deeper in overbalanced conditions. However for underbalanced conditions, although the wellbore pressure is less than the reservoir pressure, water still invades the matrix rock due to the strong capillary suction and causes an increase in water saturation around the wellbore. Thus, damage caused by water blocking might still be significant even in the case of underbalanced drilling in tight formations, owing to the ability of high and negative capillary pressure (water suction) to compensate for relatively low mud pressure in the common case where the tight gas formation is strongly water wet.

3.1.5 Phase Trapping Caused by Oil Based and Water Based Drilling Fluids

The simulation model is run to evaluate the effect of water and oil invasion damage on well productivity. To evaluate phase-trap damage, the model is run for the cases of no liquid invasion prior to gas production (no damage), injection of water into the model, followed by gas production (water damage), and injection of oil into the model, followed by gas production (oil damage).

The simulation results for cumulative gas production rate are shown in Fig. 3.7, which indicate that the well productivity is reduced in both oil and water invasion cases due to the liquid phase trapping that can not be removed by gas production. However, the well productivity is more sensitive to water invasion damage than invasion of oil, and in the case of oil invasion, the damaging effect is significantly less than water invasion.

3.1.6 Water Blocking Damage in Hydraulically Fractured Wells

Hydraulic fractures are introduced to the reservoir scale simulation model by defining high permeability planes perpendicular to the wellbore. The fractured model is run for several cases as detailed in Table 3.1. The models A1 to A6 refer to hydraulic fractured models with no water invasion damage, and models B1 to B6 refer to hydraulic fractured models with no water invasion damage. The models are run to understand the effect of initial water saturation and water invasion damage on gas production rate in non-fractured and hydraulically fractured wells.

The simulation results for the effect of initial water saturation are shown in Fig. 3.8, which indicate significant effect of S_{wi} on well productivity. For all the cases, sub-normal S_{wi} provided significantly higher gas production rate.

Fig. 3.7 Effect of water and oil invasion damage on gas recovery

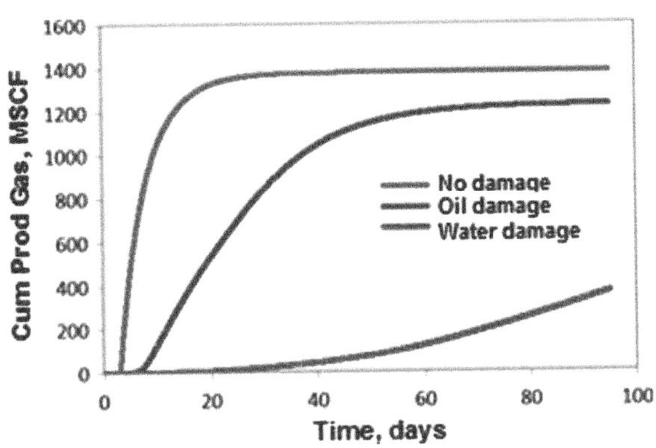

Table 3.1 Simulation results for injected water and recovered water

Simulation results for water production/injection	Scenarios	Cumulative Injected Water (bbl)	Cumulative Produced Water (bbl)
A1, normal Swi, no frac	No water invasion prior to gas production	–	–
A2, normal Swi, 1 frac			
A3, normal Swi, 5 fracs			
B1, sub-normal Swi, no frac			
B2, sub-normal Swi, 1 frac			
B3, sub-normal Swi, 5 fracs			
A4, normal Swi, no frac	2000 bbl/d water injection prior to gas production	1829	829
A5, normal Swi, 1 frac		1872	911
A6, normal Swi, 5 fracs		2046	1164
B4, sub-normal Swi, no frac		4443	134
B5, sub-normal Swi, 1 frac		4472	146
B6, sub-normal Swi, 5 fracs		4600	192

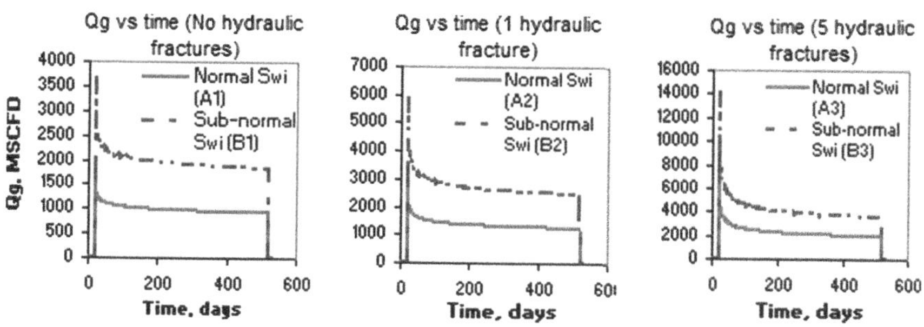

Fig. 3.8 Effect of initial water saturation on gas production rate

The simulation results in Fig. 3.9 show the effect of water blocking damage in tight formations with normal S_{wi}. In the case of non-fractured well, water blocking damage causes significant drop in gas production rate, and in the case of a fractured well, the hydraulic fractures could improve well productivity. With 5 hydraulic fractures, the stabilized gas production rate at late time is almost similar in the cases of damaged and non-damaged wells (A3 and A6), which indicates that the dominant effect of large hydraulic fractures compared with formation damage effect.

The summary of simulated results for cumulative injected water during leak-off, and cumulative produced water during clean-up and gas production are reported in Table 3.1, which indicate that in the reservoirs with sub-normal S_{wi}, most of the injected water during hydraulic fracturing is held inside the reservoir rock by capillary imbibition.

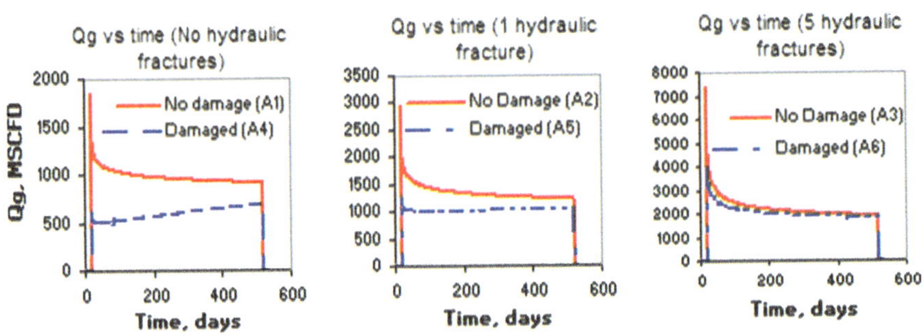

Fig. 3.9 Effect of water invasion damage on gas production rate

Compared with the normal S_{wi}, the sub-normal S_{wi} models have larger leak-off of liquid into the formation, and significantly smaller volume of cumulative water produced back. In other words, water phase trapping damage is more significant in tight gas reservoirs that have sub-normal initial water saturation.

The simulation results in Fig. 3.10 show the significance of damage control for well productivity improvement are shown. In the case of normal S_{wi}, cumulative produced gas from the well with a single hydraulic fracture that is damaged by water invasion (A−5) is not significantly different as compared with the well with no hydraulic fractures and no damage (A−1). In other words, in the case of single hydraulic fracturing, the well productivity may not be improved noticeably if water blocking damage is significant. For both damaged and non-damaged formation, the models with five hydraulic fractures provided significantly better productivity.

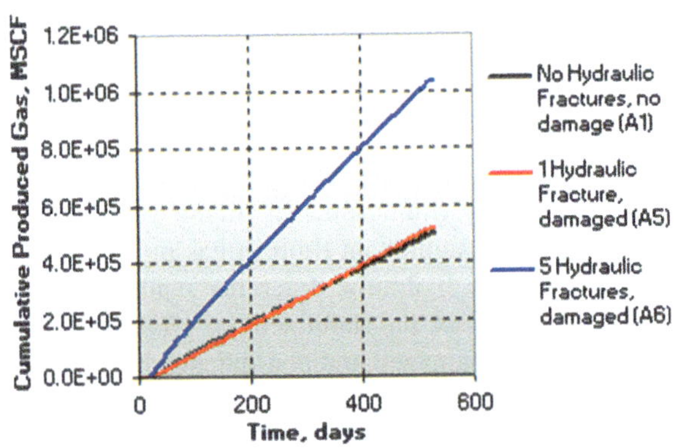

Fig. 3.10 Effect of water invasion on productivity of hydraulic fractured wells

3.1.7 Damage to Natural Fractures

The reservoir scale simulation model is run in the cases where the well intersects no natural fractures, the well intersects open natural fractures, and the well intersects the natural fractures that have been plugged at the well location. The simulation results for gas production are shown in Fig. 3.11, which indicate a significant reduction in cumulative gas production in the case that the natural fractures are damaged (well productivity close to a non-fractured reservoir).

3.2 Effect of Wellbore Related Parameters on Well Productivity

The simulation models are built based on the available field data from the West Australian tight gas reservoir to understand how different parameters affect well productivity in tight gas reservoirs.

3.2.1 Hydraulic Fractures

Hydraulic fractures may propagate differently in tight formations depending on wellbore direction and stress anisotropy [1, 2]. In a reservoir with normal stress, depending on wellbore direction and in situ stresses, the hydraulic fractures might be parallel to the wellbore (longitudinal), or perpendicular to the wellbore (transverse).

Simulation models are built for the cases of a non-fractured and a multi-fractured well, as open-hole (or a fully-perforated cased hole), considering all the hydraulic fractures have similar size (75 ft half length size and 100 md ft conductivity). The

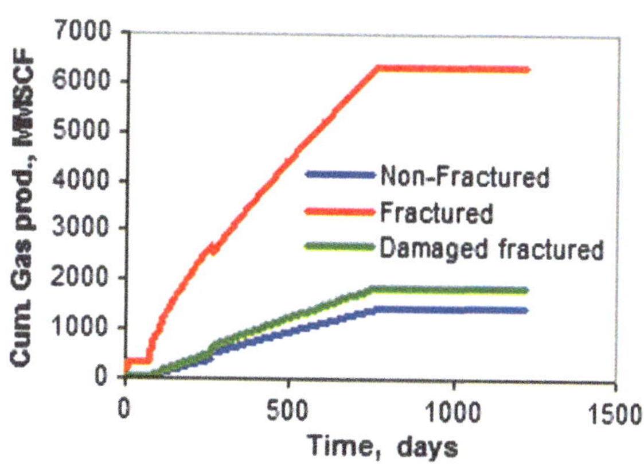

Fig. 3.11 Effect of damage to natural fractures on productivity

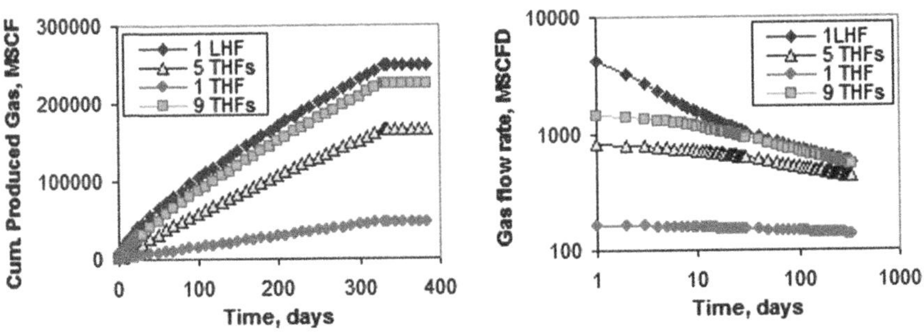

Fig. 3.12 Productivity of hydraulic fractures: transverse versus longitudinal

following cases are run in order to understand the effect of wellbore direction on productivity of hydraulically fractured horizontal well: No hydraulic fractures (No HF), one longitudinal hydraulic fracture along wellbore (1 LHF), one transverse hydraulic fracture perpendicular to wellbore (1 THF), five transverse hydraulic fractures perpendicular to wellbore (5 THFs), and nine transverse hydraulic fractures perpendicular to wellbore (9 THFs).

The simulation results are shown in Fig. 3.12, which indicate that the longitudinal hydraulic fracture provides significantly higher gas production rate compared with the single, 5 or 9 transverse hydraulic fractures. Although each single transverse hydraulic fracture has similar size (volume) compared with the longitudinal hydraulic fracture, since a hydraulic fracture along wellbore has larger direct contact area to the wellbore, it provides higher gas rate compared with the transverse hydraulic fractures perpendicular to wellbore.

3.2.2 Drilling Direction and Permeability Anisotropy

A reservoir simulation model is used to understand how the direction of horizontal drilling can affect well productivity in tight gas reservoirs that have significant horizontal permeability anisotropy [1, 2]. The horizontal well length is 1250 ft in a tight formation with significant horizontal permeability anisotropy of 5 ($K_{h,max}/K_{h,min} = 5$). This level of anisotropy could be produced, for example, by oriented sand bodies or channels. First, the wellbore direction is considered to be perpendicular to the direction where permeability is larger. Then the model is run considering the wellbore direction perpendicular to the direction where permeability is the minimum.

The results for gas production rate from the model are shown in Fig. 3.13 [1, 2]. The horizontal wells drilled in the direction perpendicular to the direction of maximum permeability (drilling in the direction of minimum horizontal stress, if permeability anisotropy is caused by stress anisotropy) may provide noticeably higher gas production rate compared to the wells that are drilled perpendicular to

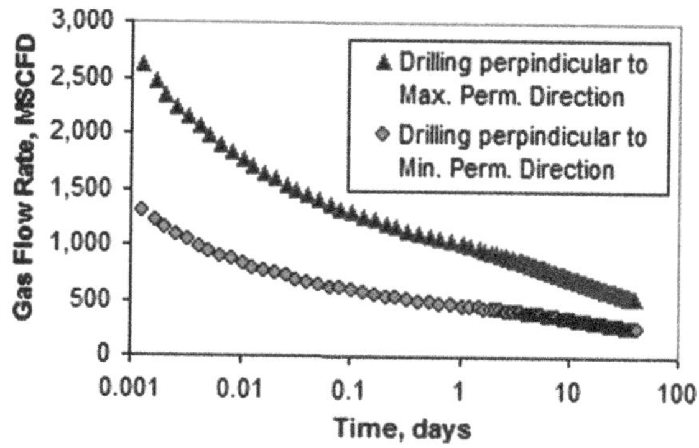

Fig. 3.13 Effect of wellbore direction on well productivity

the direction of minimum permeability (drilling in the direction of maximum horizontal stress, if permeability anisotropy is caused by stress anisotropy).

3.2.3 Wellbore Breakouts

In order to understand the effect of wellbore break out on well productivity, the horizontal well model is run for a zero skin cased-hole perforated horizontal well (wellbore diameter of 8 inches), and zero skin open-hole horizontal well with enlarged wellbore (wellbore diameter of 20 inches).

The production predictions from the models as shown in Fig. 3.14, they indicate that open-hole completion in the gas wells with large wellbore breakouts can provide significantly higher initial gas production rate compared with cased-hole completion system. In use of open-hole completion, the enlarged wellbore due to break outs can result in higher effective wellbore radius, and therefore a lower skin factor and higher productivity [3].

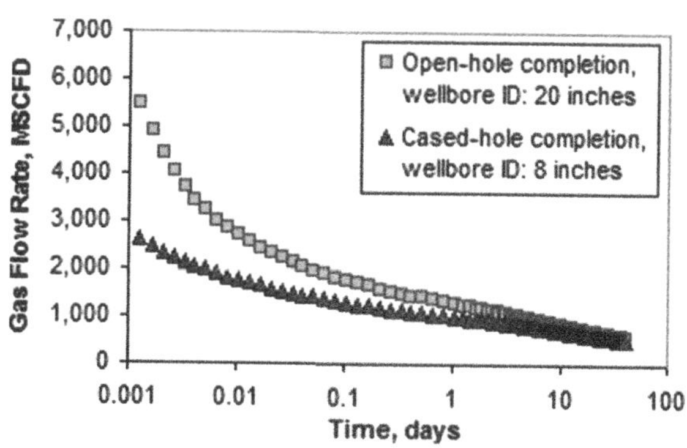

Fig. 3.14 Effect of wellbore breakouts on well productivity

3.2.4 Perforation Parameters

Perforation parameters can control skin factor and well productivity. To understand the effects, a model was built and run for damaged and non-damaged perforated cores in the cases of tight and conventional reservoirs. The simulation results show that the perforation tunnel provides improvement in core flow efficiency, which is more noticeable in tight sand cores compared to conventional cores as shown in Fig. 3.15.

The simulation model is also run for the case of an open-hole perforated damaged tight gas reservoir. The results as shown in Fig. 3.16 indicate that if the damaged zone is not fully bypassed by the perforations, the flow efficiency is still significantly reduced. The flow efficiency is sensitive to perforation parameters and highlights the importance of passing damaged zone radius, especially in tight gas reservoirs. According to the simulation results, even for open-hole wells in tight formations, improved productivity (flow efficiency greater than 1) may be achieved by creating deep, clean perforation tunnels that can bypass the mechanical damaged zone.

3.2.5 Liquid Loading in Wellbore

Well stimulation and fracturing operations in tight formations cause significant liquid leak-off into the reservoir rock. The invaded liquids when produced, they may be loaded in wellbore during post-fracturing clean-up period, since natural gas flow rate may not be high enough to lift the wellbore liquids to surface. A series of simulation runs are carried out to model the wellbore phenomena for a horizontal deviated wellbore. The results as shown in Fig. 3.17 indicated water loading problem in wellbore at 4 MMSCFD gas production rate. The problem becomes more serious when gas flow rate is reduced to 1 MMSCFD. Based on the simulation results, liquid loading can be one of the main causes of low well productivity in tight gas wells.

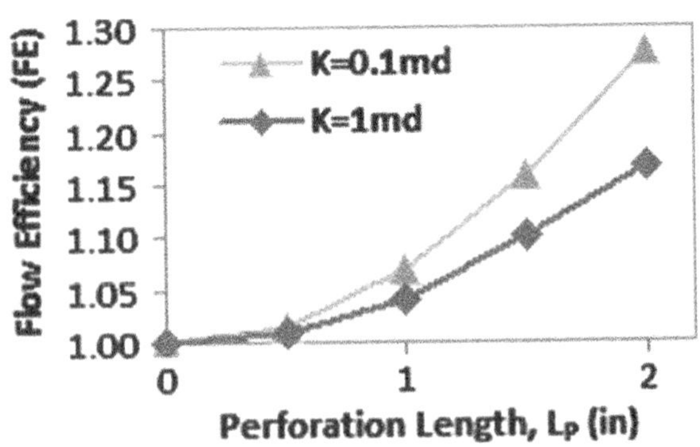

Fig. 3.15 Effect of perforation tunnel length on flow efficiency

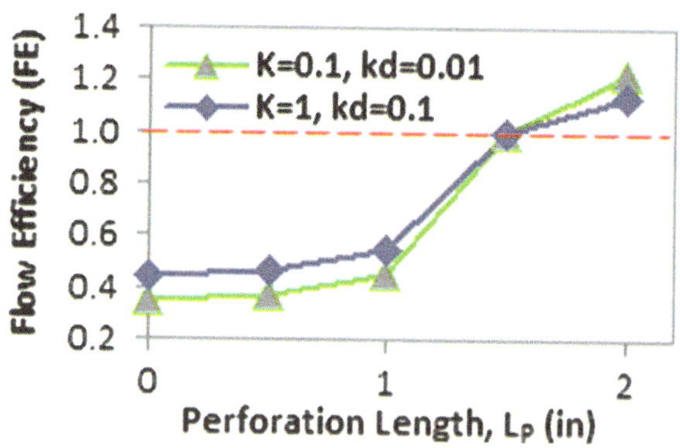

Fig. 3.16 Effect of perforation tunnel length on flow efficiency

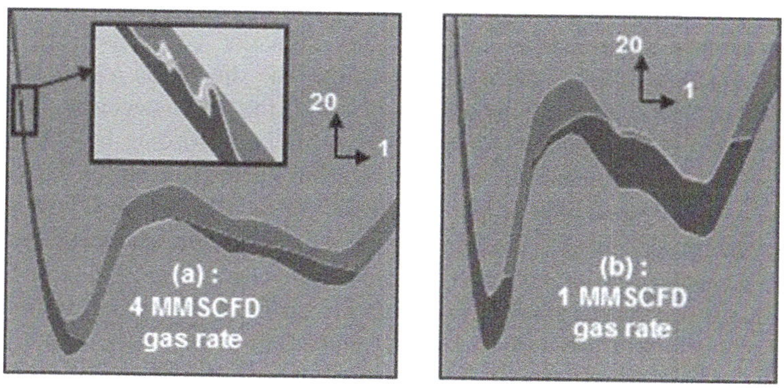

Fig. 3.17 Simulation of liquid loading in wellbore (effect of gas production rate)

The well production performance modelling results also showed that use of oil based mud instead of water based mud can help reducing liquid loading, since oil has less density than water, and therefore gas can better lift the liquid to surface. Therefore, underbalanced drilling using non-aqueous liquid can reduce the issues related to liquid loading in wellbore during clean-up. Tight gas well productivity can also be further improved by gas lift and optimizing the producing liner and tubing size.

3.3 Effect of Natural Fractures Parameters on Welltest Response

A tight gas reservoir model is built as a dual-porosity dual-permeability medium using the Kazemi model with parallel layers of low permeability tight matrix and high permeability fractures. The model generates transient pressure data for gas

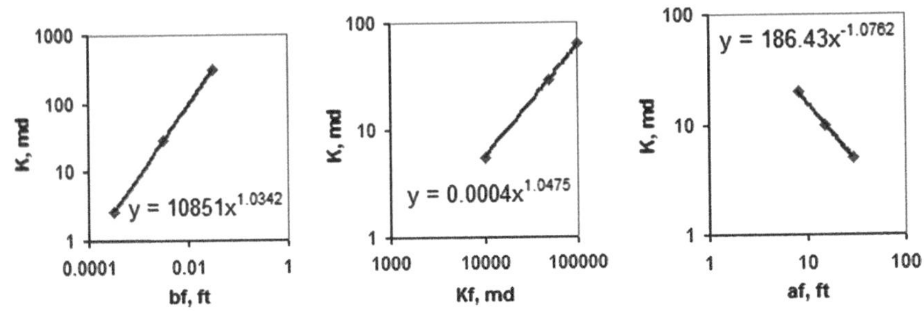

Fig. 3.18 Relationship between fracture parameters and welltest permeability

production and pressure build-up periods to show the relationship between welltest permeability and natural fractures parameters.

The model is run for different values of fracture aperture, fracture permeability, matrix permeability, matrix compressibility, fracture compressibility, matrix porosity, fracture porosity and fracture spacing. The simulation results for sensitivity of welltest permeability indicated that among the parameters examined, only the fracture aperture, fracture permeability and fracture spacing have significant impact on welltest permeability (k), and the other parameters can be disregarded. According to the simulation outputs as shown in Fig. 3.18, welltest permeability, k, is directly proportional to fracture aperture, b_f, and fracture permeability, k_f (the power exponent of +1 approximately), and has inverse relationship with fracture spacing, a_f (the power exponent of −1 approximately). The observations are in good agreement with the derived Eqs. 2.8 and 2.9 regarding how the parameters control fracture permeability, which confirms the reliability of the proposed equation.

The curve fitting functions and multi-variable regression on the data resulted in the following correlation in field units for fracture permeability estimation in the tight gas reservoir:

$$K_f = 0.795 * K_{welltest} * \left(\frac{a_f}{b_f}\right)^{1.04} \quad (3.1)$$

3.4 Pressure Measurement in Tight Gas Reservoirs

Measurement of formation pressure in tight gas reservoirs using formation testers may be affected by the drilling fluids invasion into the formation, and to understand the effect, the reservoir simulation model is run for formation testing at different depths using 20 cc/min production of reservoir fluid, followed by pressure build-up.

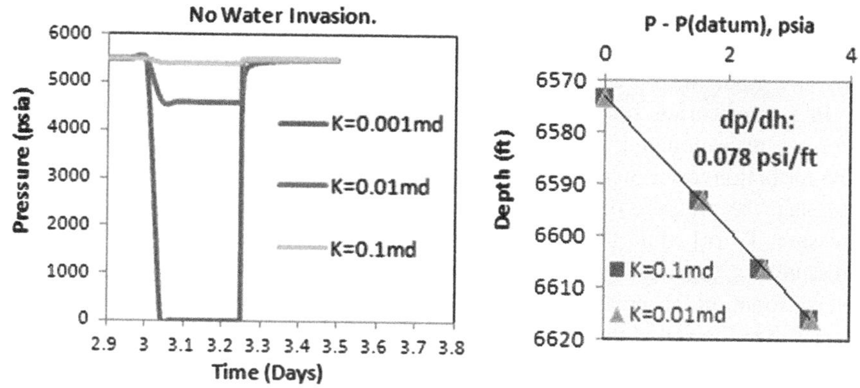

Fig. 3.19 Formation testing response in the case of no water invasion

The simulation model is first run for a non-damaged rock (no liquid invasion). The results as shown in Fig. 3.19, they indicate that in the case of very low permeability of 0.001 md, pressure around the tested interval drops to zero (the tight rock cannot provide the flow rate), resulting in a dry test. In the case of higher permeability, the test is normal. For the normal tested points, the plot of pressure difference ($P–P_{datum}$) versus depth resulted in gas gradient (dp/dh of 0.078 psi/ft). It can be concluded that for a gas bearing zone drilled with air (similarly, for a water bearing zone in a water-wet formation drilled with water-based mud), the tool measures the true formation pressure and provides a reliable pressure gradient, since there is no capillary pressure effect between the invading fluid and the formation fluid.

In the case of water invasion from wellbore into the reservoir, the presence of water filtrate in the invaded radius of a gas zone may affect pressure measurement during formation testing. The simulation outputs for pressure versus depth are shown in Fig. 3.20. The results indicate that in the good permeability zone (0.1 md), the invasion of water does not have significant impact on the pressure

Fig. 3.20 Effect of liquid invasion on formation testing response

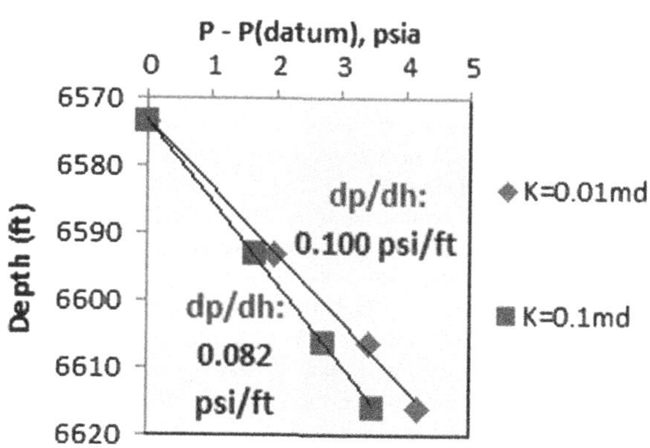

gradient measured by the formation testing tool (0.082 psi/ft). However in tighter sections (permeability of 0.01 md), the water invasion effect causes increase of pressure gradient to 0.1 psi/ft.

In a gas bearing zone drilled with water-based mud (or oil-based mud), the pressure measurements in the gas zone are influenced by liquid invasion effect. The formation tester measures the pressure of filtrate in the invaded zone, which is less than the pressure of the reservoir fluid (gas) by the amount of the capillary pressure, P_c [4], thus the tool under-estimates the value of the true formation pressure ($P_{measured} < P_{actual}$). In addition, presence of water in the gas zone causes over-estimation of pressure gradient measured by the tool (pressure gradient higher than the actual gradient), according to the simulation results. These effects (under-estimated reservoir pressure and over-estimated pressure gradient) overall may cause the gas–water contact depth that is based on uncorrected field measurements to be different than the actual gas–water contact depth. A reliable estimation of formation pressure, gradient of pressure and gas–water contact depth for a tight gas reservoir requires that we understand these processes, and take into account the capillary pressure and liquid invasion effects.

3.5 Summary

In this section, the effect of various well and reservoir parameters on well productivity was studied using reservoir simulation models based on the tight gas field data: phase trapping damage in non-fractured and hydraulically wells, in under-balanced and over-balanced drilling, and in the case of invasion of different fluids into the tight formation (water and oil). The effects of permeability anisotropy, wellbore breakouts, perforation parameters, liquid loading in wellbore, different hydraulic fractures systems and natural fractures on well productivity are presented, and it was also shown how formation testing pressure measurements and gas water contact determination may be influenced by liquid invasion into tight gas reservoirs.

References

1. Bahrami H, Rezaee R, Asadi M, Minaeian V (2013) Effect of stress anisotropy on well productivity in unconventional gas reservoirs. In: SPE 164521, SPE Production and Operations Symposium, Oklahoma
2. Bahrami H, Mohemad Nour J, Alwerfaly Kh, Mutton G, Owais SA, Al-Waley A (2013) Whicher range field development planning. In: Study project in FDP course (MSc level, semester-3), Curtin University, Perth
3. Bahrami H, Rezaee R, Nazhat D, Ostojic J (2011) Evaluation of damage mechanisms and skin factor in tight gas reservoirs. APPEA Journal, Canada
4. Elshahavi H, Fathy K, Heikal S (1999) Capillary pressure and rock wettability effects on wireline formation tester measurements. In: SPE 56712, SPE Annual Technical Conference and Exhibition, Houston, Texas

Chapter 4
Tight Gas Field Example: Effect of Damage Mechanims on Well Productivity

Whicher Range (WR) field, located in the Southern Perth Basin, is a large, low permeability and very heterogeneous gas reservoir. It consists of stacked and isolated lenses of sand separated by shale layers. There are five wells in WR gas field, which are drilled mainly as vertical well using water based mud in over-balanced conditions, and completed as cased-hole perforated. Figure 4.1 shows location of the WR wells. Some faults in this field have divided the reservoir into different compartments. The wells WR 1, 2, 4 and 5 are drilled in the west reservoir compartment, and WR 3 was drilled in the east reservoir compartment. Total gas in place estimations range from 1.5 to 5 TCF for the field [1].

Hydraulic fracturing was performed in WR-1, WR-3 and WR-4 using water based fracturing fluid. None of the wells produced at viable rates despite various attempts to stimulate the formation [1].

There is a long history of various DST and production tests performed in this field. Despite well stimulation and other well intervention operations, gas production rates are found to be low and did not meet the expectations. In this chapter, various categories of data in WR field are reviewed and analysed in order to obtain a better understanding of the reservoir, to evaluates possible mechanisms that might have contributed to the low productivity and assess the feasibility of achieving commercial production rates.

4.1 Wellbore Instability

The wells drilled in the tight sand formation had severe wellbore instability issues during drilling, which caused enlargement of wellbore up to 20–25 inches (2–3 times bigger than the bit size) across majority of the tight sand intervals, as shown in Fig. 4.2.

N. Bahrami, *Evaluating Factors Controlling Damage and Productivity in Tight Gas Reservoirs*, Springer Theses, DOI: 10.1007/978-3-319-02481-3_4,

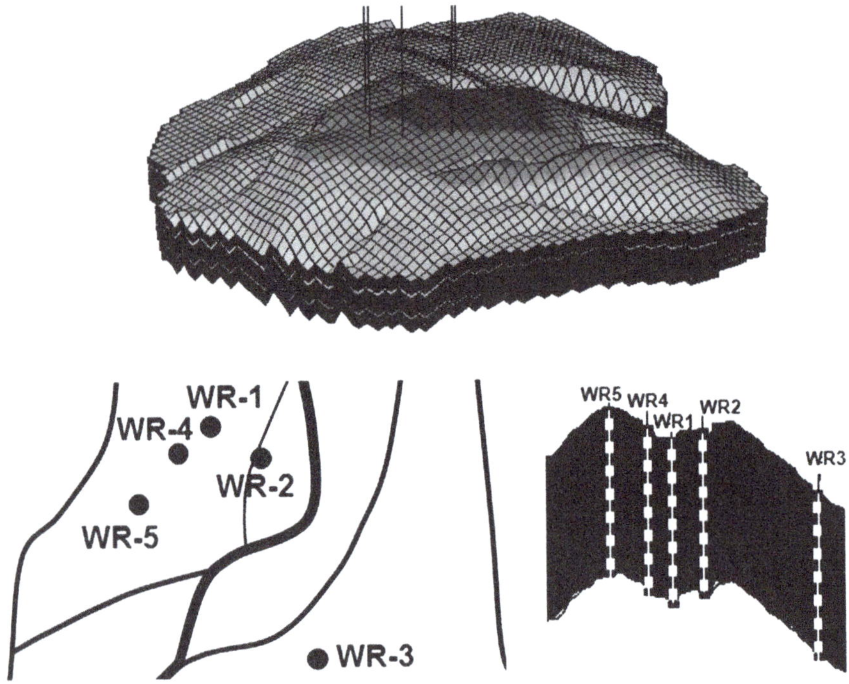

Fig. 4.1 Whicher Range wells

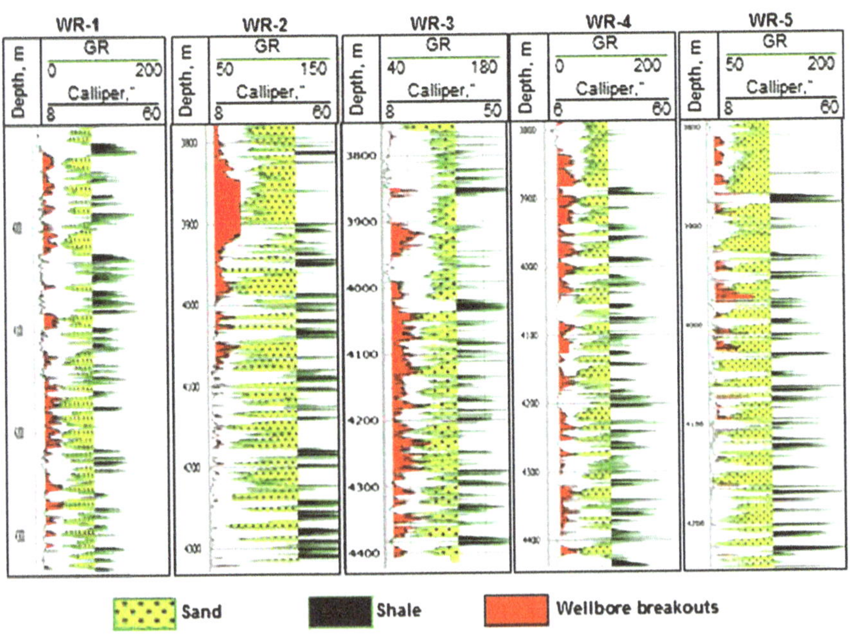

Fig. 4.2 Wellbore instability in WR tight sand zones [2, 3]

4.2 Perforation Data

Perforation jobs in these wells are mainly performed using 2 1/8″ EJ guns (API RP19B standard penetration of 23″). According to the well data, the casing has 7″ internal diameter and the borehole may have 10″ diameter in the direction of maximum stress where the borehole is stable, and 20″ in the direction of minimum stress due to the wellbore breakouts. Damaged zone radius is also assumed as 5″.

Schlumberger perforation analysis software (SPAN 7.02) is run for the tight formation to analyse the perforation data in different directions. The results for perforation jet penetration and skin factor as presented in Table 4.1 and Fig. 4.3, they indicate a positive skin factor for the tunnels, since penetration is not deep enough to efficiently connect the wellbore to the formation virgin zone. In fact, the perforation is less efficient in the direction of $\sigma_{\min}$ due to the large cement volume behind the casing in the intervals with enlarged wellbore.

Because of the reservoir rock tightness and also presence of the large wellbore breakouts behind casing filled by cement, the perforation penetration is significantly reduced. The perforation jet penetration into the tight formation is not deep enough, and therefore the wellbore may not have effectively been connected to the undamaged reservoir rock. The poor perforation efficiency might be the primary reason of the low productivity in the cased-hole perforated well.

Table 4.1 Model predictions for perforation jet penetration and skin

Casing	Gun type	Perforation tunnel direction	Formation penetration, inches	Perforation skin
7″	2 1/8″ gun	σmax	6.0	2.3
		σmin	4.2	11.5

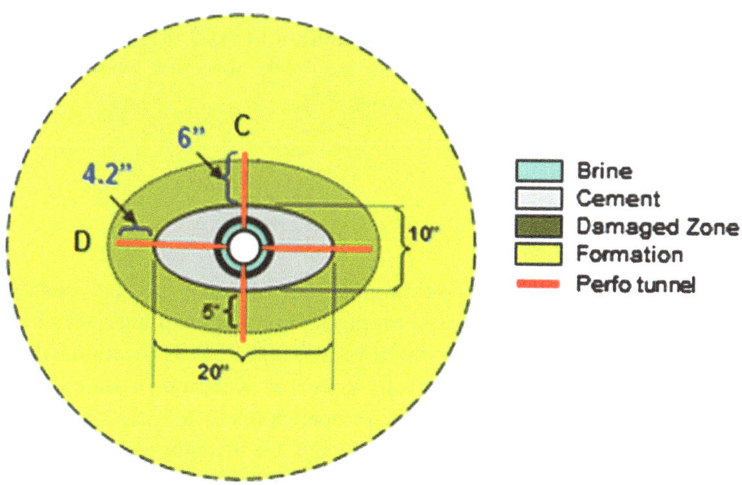

Fig. 4.3 Effect of wellbore breakouts on perforation performance (2 1/8″ EJ)

Table 4.2 Model predictions for perforation jet penetration and skin

Casing	Gun type	Perforation tunnel direction	Formation penetration, inches	Perforation skin
7″	4 1/2″ gun	Max	13.3	−0.6
		Min	11.5	0.1

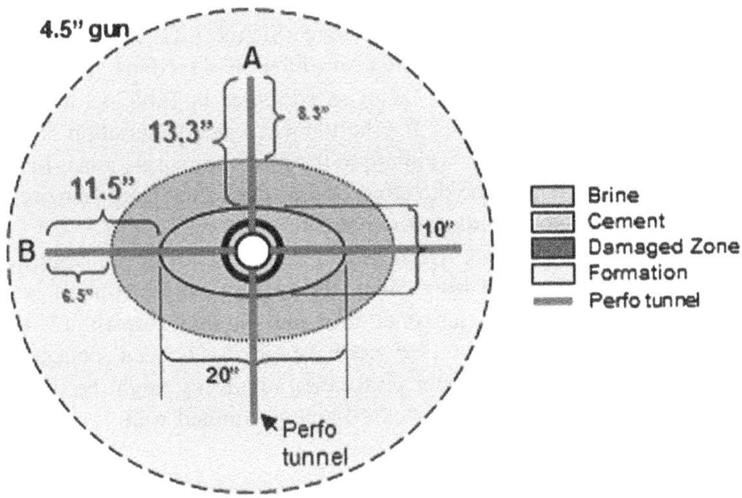

Fig. 4.4 Effect of wellbore breakouts on perforation performance (4.5″ HSD PJO)

In order to have an improved well productivity for a cased-hole perforated well completion system, the optimum option as shown in Table 4.2 and Fig. 4.4, might be to use deep penetrating perforation charges (such as 4 ½″ Power-Jet Omega HSD guns, with API RP19B standard penetration of 60″) oriented towards maximum stress direction i.e., 180 degree phasing, in the direction where the wellbore is stable and has no considerable breakouts [4].

4.3 Reservoir Fluid

Whicher Range tight gas reservoir produces mainly dry gas and contains low amounts of heavy components based on production data and analysis of the PVT fluid samples as shown in Fig. 4.5 (GOR of 200,000 SCF/STB estimated the Dew Point pressure of 2600 psia using Peng Robinson equation of state). Therefore, condensate banking and partial blockage of open pores by the condensate drop outs in the formation near the wellbore would not be a major issue in WR wells. However at very low flowing bottom-hole pressure (very large pressure drawdown), condensate may drop out of the gas phase in the formation and reduce the well

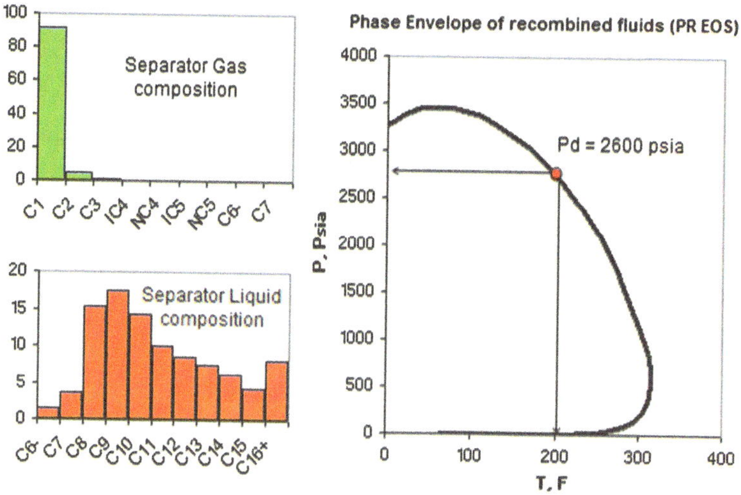

Fig. 4.5 Composition (%) of reservoir fluid components in WR field

productivity due to the condensate phase trapping damage. Different fluid samples taken from different zones of WR-1 and WR-4 showed almost similar fluid compositions, indicating that the wells are in the same reservoir compartment.

4.4 Initial Water Saturation

Estimation of water saturation using petrophysical log data in WR wells have high uncertainties due to the formation tightness and significantly large wellbore breakouts across majority of the sand intervals. As shown in Fig. 4.6, there is no agreement between the petrophysical evaluation results of the two different data service companies that worked on the log data to estimate water saturation. Therefore by using the log data, it is not feasible to comment confidently on initial water saturation or have evaluations regarding depth of water invasion into the formation during drilling. Even porosity estimation from log data may have uncertainties and overburden core porosity data may be more reliable to be used in the reservoir studies. Overall, petrophysical evaluations in this field indicate relatively high initial water saturation, $S_{w,average}$ of around 60 % (low initial gas saturation) in the reservoir. Also, the irreducible water saturation is believed to be high due to extensively developed micro-porosity.

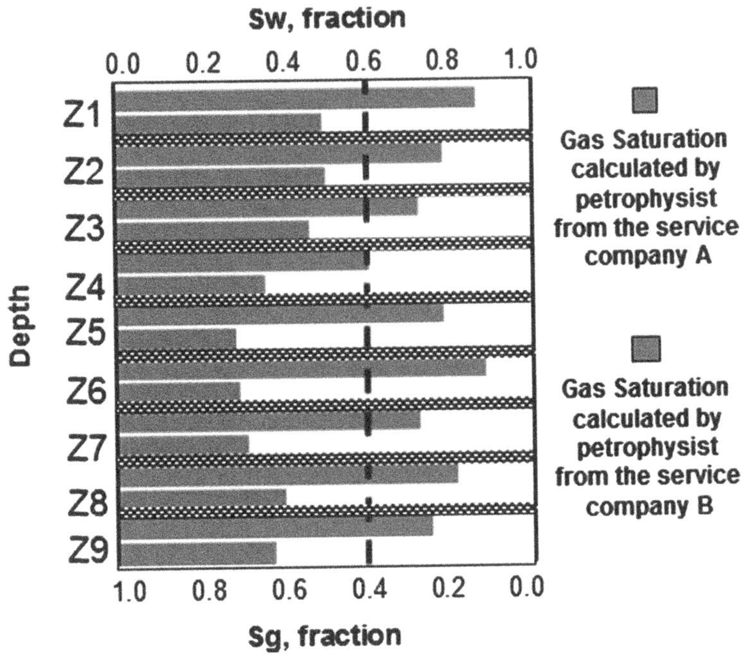

Fig. 4.6 Water saturation variations along wellbore in the tight sand formation in Perth Basin (Total net thickness of the porous sand intervals: 370 ft)

4.5 Relative Permeability and Capillary Pressure

There are limited core data available in the field, and they are studied to evaluate production performance of the tight gas reservoir. Overburden core porosity and core permeability for the core samples that have core analysis tests data available are shown in Fig. 4.7. Among the tested core samples, the good quality core sample was used to have reliable core flood experiment data measured. The core sample was taken from the zone Z3 shown in Fig. 4.3, and it has overburden permeability and porosity of 0.035 md and 9.6 % respectively.

The core flooding experiment was performed under the conditions of 80°C, confining pressure of 5000 psia, constant outlet pressure, and constant injection rate of 5 cc/hr. The core sample was first fully saturated with water (salinity of 20000 ppm), and then gas (Methane) was injected into the core sample. The pressure drop across the core sample and water production rate were measured during core flooding (first drainage), until the irreducible water saturation and stabilized conditions reached. The core flood experiment showed the end point gas relative permeability of 0.35. After that, brine was injected into the core sample until the critical gas saturation and stabilized conditions reached. Then gas was injected into the water-saturated core sample again (secondary drainage), that resulted in the end point gas relative permeability of 0.25. The relative

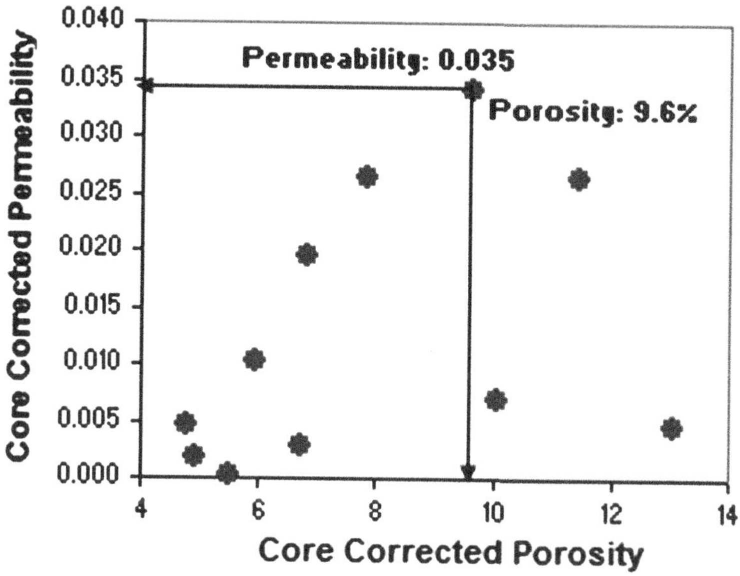

Fig. 4.7 Corrected permeability and porosity for the core samples of the tight sand formation, which have core analysis tests available

permeability in the second drainage is lower compared to the first drainage, indicating damage to the core permeability caused by brine injection.

The core flooding data were analysed using Sendra core flooding simulation program, the test history data were history matched, and the relative permeabilities were calculated accordingly. Figure 4.8 shows the lab facilities of Petroleum Engineering Department at Curtin University, the core flooding data for pressure drop and water production rate in the first drainage, the match of test history, and the calculated relative permeability curves. The results indicate that the relative permeability to water is sig-nificantly lower compared with the relative permeability to gas (typical behaviourin water sensitive formations). The core flooding experiment also indicated the irreducible water saturation of 60 %. For the core sample, air-mercury capillary pressure data are also analysed and using some conversion factors, the gas–water capillary pressure data could be determined.

4.6 Hydraulic Fracturing

Hydraulic fracturing in the tight gas wells had 3–4 fracturing stages, with average maximum pumping pressure of 10,000 psia, average pump rate of 20 bbl/min, average total proppant per stage of 35,000 lbs, and average total liquid injected per stage of 2,500 bbls. The fracturing models predicted 100–150 ft fracture half length size, and fracture conductivity of 700–2100 md.ft for the fractures.

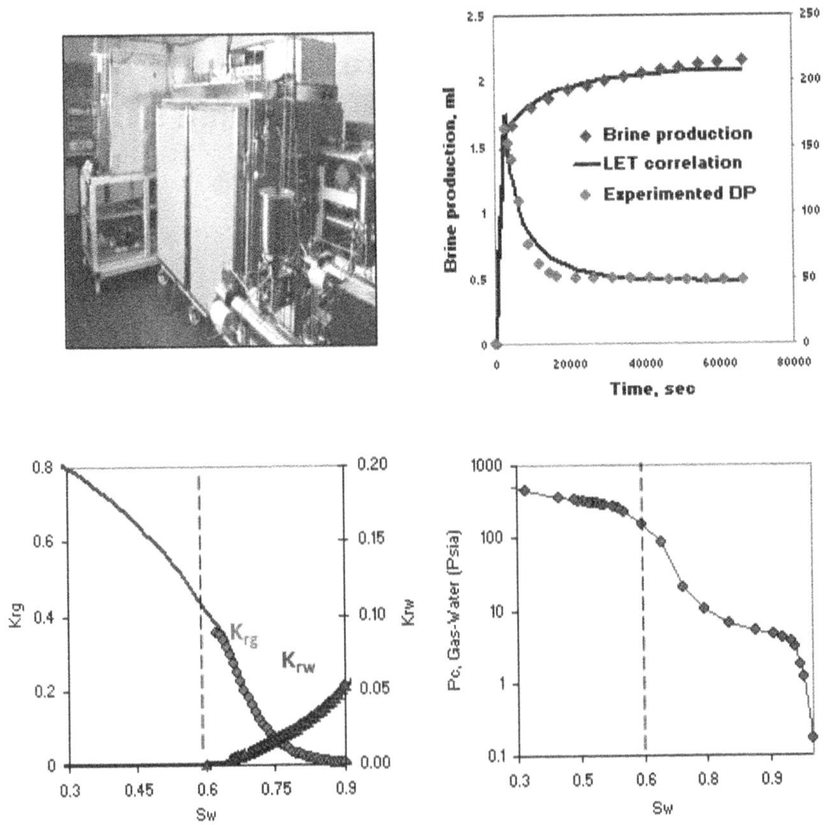

Fig. 4.8 The relative permeability curves for the tight sand core sample

After the hydraulic fracturing operations and then cleaning up the well from the fracturing liquids by gas production, only 10–60 % of the injected fracturing liquid could be recovered (40–90 % of fracturing liquid being trapped in the reservoir), which may have caused significant damage to near wellbore formation.

4.7 Welltest Analysis

Pressure build-up test data in one of the hydraulically fractured vertical wells in this field (with longitudinal fractures) are analysed in order to estimate reservoir permeability, hydraulic fracture size, and evaluate the well productivity. As shown in Fig. 4.9a, the test duration was not long enough to reach radial flow regime, and analysis of the welltest data have uncertainties using conventional methods.

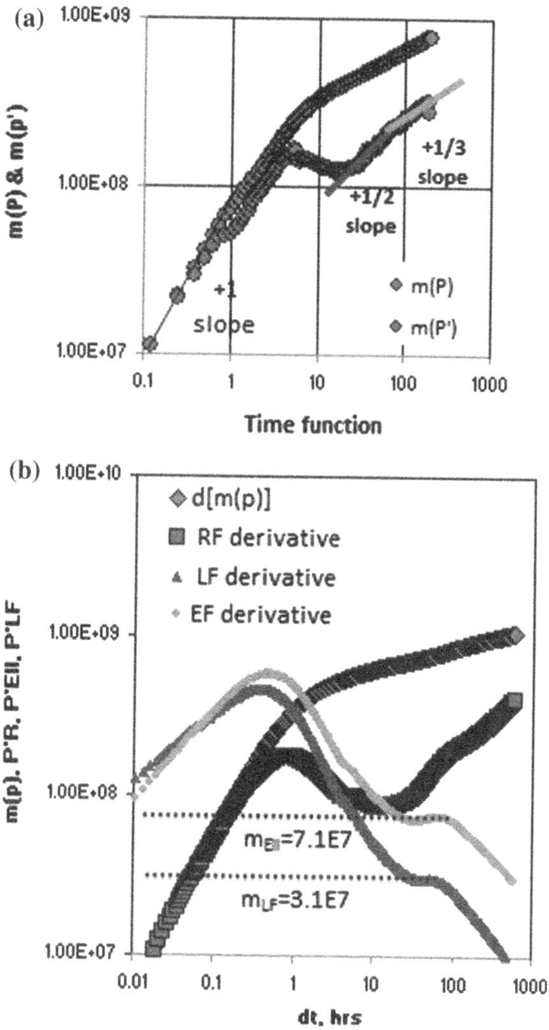

Fig. 4.9 Welltest analysis in the tight gas well

Using the diagnostic plot, linear and elliptical flow regimes are both detected, and therefore the new proposed method of welltest analysis was used based on Linear Flow (LF) derivative and Elliptical Flow (EF) derivative to analyse the data. By integrating the diffusivity equation solution for the two reservoir flow regimes, the two unknowns K and X_f can be calculated (Eqs. 2.1 and 2.2): reservoir permeability of 0.002 mD and hydraulic fracture half-length of 160 ft based on the total sand thickness of 800ft (Fig. 4.9b). The results arc in good agreement with the expected hydraulic fracture size based on the fracturing job design (100–150 ft).

4.8 Formation Sensitivity to Water Invasion Damage

The core samples in the tight gas field that are tested and analyzed using X-Ray diffraction, they detected smectite, meaning that the tight formation can have medium to strong sensitivity to fresh or sodium chloride waters (indicating a water sensitive tight formation). With such minerals present, clay swelling may be reduced by using relatively high concentrations of KCl, sometimes accompanied by $CaCl_2$ in the drilling and stimulation fluids.

4.9 Well Production History (Pre-Frac and Post-Frac)

The well after completion and perforation did not produce at economical rates, and it was later hydraulically fractured in multi-stages using KCl based fracturing fluid in order to improve well productivity. The post fracturing production rates are shown in Fig. 4.10, that indicates the stabilized gas production rate of 1.1 MMSCFD after the well clean-up. The results does not show significant production improvement compared to pre-fracturing flow rate. In the fracturing job, approximately 60 % of the fluids were recovered during post fracturing production test, and significant amounts of water based treating fluids was trapped in the water-sensitive tight formation (we believe that this trapped in the reservoir,

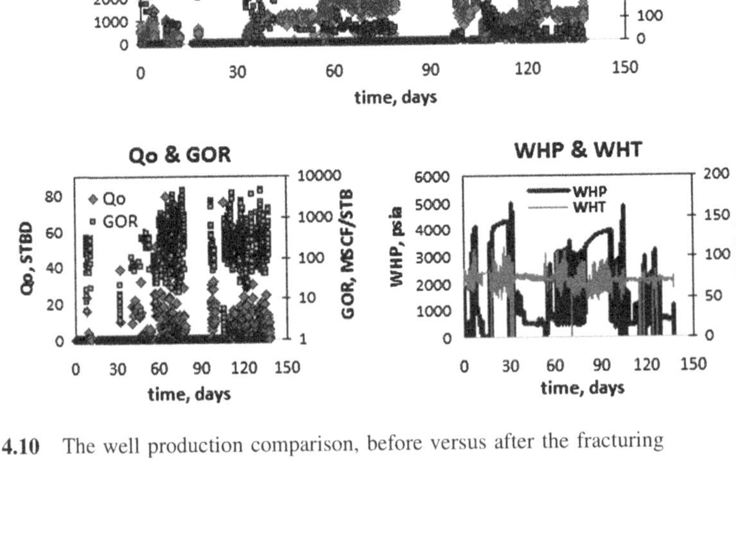

Fig. 4.10 The well production comparison, before versus after the fracturing

causing significant damage to reservoir permeability). Therefore it could be concluded that the leak-off of water into the tight sand gas reservoir during fracturing might be the reason for the low well productivity after stimulation.

In term of hydraulic fractures performance, the fracturing job was not successful. The well may produce gas at a commercial rate using massive hydraulic fracturing and creating large treated zone volume, which might be achieved by use of non-damaging fracturing liquid, increasing numbers of fracturing stages (For instance 10–20 fracturing stages instead of 3–4), higher pumping rates of fracturing fluid (For instance 60–80 bbl/min instead of 20 bbl/min) and larger volumes of liquid and proppant injection per stage.

4.10 Formation Damage Caused by Water and Oil Invasion

Laboratory experiments on the West Australian core samples are performed using the core flooding facilities in Petroleum Engineering Department at Curtin University, in order to compare damage to the core permeability, caused by water invasion versus oil invasion.

Absolute permeability of each core sample is measured at 100 % gas saturation. The characteristics of the core samples tested for damage evaluation are reported in Table 4.3. The test conditions are 5,800 psia pore pressure and 109 °C temperature.

4.10.1 Damage Caused by Water Invasion

For water damage evaluation, the core samples are saturated with water, followed by flooding of gas into the core samples until water saturation is reduced to the minimum of residual water saturation, and then core effective permeability to gas is measured (K_{rg} @ S_{wr}). From the core experiments as reported in Table 4.4, water invasion causes the relative permeability of gas to be reduced to 0.22–0.35, indicating around 70 % reduction in effective permeability.

Table 4.3 Details of the core samples characteristics

Core sample	Core porosity (%)	Core absolute permeability, md
Sample A1	6.7	0.003
Sample A2	9.6	0.034
Sample A3	11.4	0.026
Sample B1	4.6	0.034
Sample B2	5.4	0.032

Table 4.4 Core flooding experiment results for water invasion damage

Core sample flooded with water, cleaned up with gas			
End point methane K_r @ S_{wr}	Sample A1: 0.224	Sample A2: 0.353	Sample A3: 0.219

Table 4.5 Core flooding experiment results for oil invasion damage

Core is flooded with oil, cleaned up with gas		
End point methane K_r @ S_{or}	Sample B1: 0.507	Sample B2: 0.413

4.10.2 Damage Caused by Oil Invasion

For oil invasion damage evaluation, the core samples are saturated with oil, then gas is flooded into the core samples until oil saturation is reduced to minimum of residual oil saturation in order to measure core effective permeability to gas (K_{rg} @ S_{or}). From core experiments as reported in Table 4.5, the effective permeability is reduced by 55 % due to oil invasion.

4.10.3 Comparison of Damage Caused by Water and Oil

The experiments highlight the fact that tight gas reservoirs are subjected to invasion damage in both cases of oil filtrate invasion (from OBM) or water filtrate invasion (from WBM). In other words, even when oil-based fluids are used instead of water-based fluid for drilling or fracturing, the reservoir rock is damaged. However severity of the damage is less in the case of damage caused by oil based fluid as compared to the damage caused by water invasion.

4.11 Well Productivity Evaluation in WR Tight Gas Field

The wells in the tight gas reservoir are drilled overbalanced using water based mud, and completed as cased-hole perforated wells. Analysis of the field and lab data showed that there are various possible explanations or combination of circumstances that may have contributed to the wells' poor productivities:

- Vertical wells in low permeability gas reservoirs may not provide economical rates due to the very limited formation surface area that is open to the wellbore.
- The core data analysis (X-Ray diffraction) detected smectite which shows the reservoir rock is sensitive to water based fluids. Drilling the wells overbalanced using water based mud, may have caused significant damage and low productivity.

- In addition, perforation efficiency was low due to the reservoir rock tightness and also presence of the large wellbore breakouts behind casing filled by cement.

In the wells that are hydraulically fractured, the fracturing operations did not result in any improvement of productivity which might be due to:

- Job reports indicate that about 40 % of the water based treating fluids is not recovered. The formation is sensitive to water damage and the large leak-off of water into formation during fracturing might be one of the main factors that cause low productivity.
- Well production and test data also indicated that hydraulic fracture size is significantly smaller than the expectations. The limited size hydraulic fractures may have caused the hydraulic fractures productivity to be low.

4.12 Tight Gas Development Strategy for Optimized Productivity

Gas recovery might be low if drainage area of a well is limited to a few sand lenses (which might be the case in vertical wells drilling). The optimum strategy for the tight gas field might be to drill long highly deviated horizontal/deviated wells in underbalanced conditions using non-aqueous drilling fluid to intersect as many as possible of the sand lenses, increase the lateral reservoir exposure to wellbore, and minimize damage to formation. Drilling direction should be designed based on sand lenses width and length. Completing the well as open-hole (and running a slotted liner to control wellbore collapse) can further improve the well productivity as it provides using the advantage of enlarged wellbore and the reduced skin factor due to presence of the wellbore breakouts. Also perforating using deep perforation shots through the slotted liner and penetrating into the formation can help better bypassing the possible mechanically damaged zone in the tight formations. As perforation strategy, shooting the perforation jets aligned with the maximum stress direction (oriented perforation with 180 degrees phasing) might help improve perforation efficiency in the cased-hole perforated tight gas wells.

Improving drilling or fracturing fluid rheology and filter cake building ability, which should provide an effective cake that is later removable, it can help control the damaged zone radius and reduce the damage due to liquid invasion. Also in the high-temperature deep Whicher Range wells, thermal fracturing using cold liquid CO_2 is another method of improving the well productivity, without causing damage to the tight formation.

In the cases of significant liquid leak-off into formation, use down-hole pumps, gas lift system in the early stage (especially during cleaning the wells) and/or optimum tubing size, to unload the well from drilling or fracturing liquids, to allow the well to produce at higher gas rates.

References

1. Rezaee R, Evans B, Rasouli V, Bahrami H, Orsini C (2012) Whicher range tight gas sands study (http://www.dmp.wa.gov.au/4726.aspx)
2. Bahrami H, Rezaee R, Asadi M, Minaeian V (2013) Effect of stress anisotropy on well productivity in unconventional gas reservoirs, SPE 164521. Paper presented at the SPE production and operations symposium. Oklahoma, USA
3. Bahrami H, Mohemad Nour J, Alwerfaly Kh, Mutton G, Owais SA, Al-Waley A (2013) Whicher range field development. Curtin University, Perth, Australia
4. Bahrami H, Rezaee R, Nazhat D, Ostojic J (2011) Evaluation of damage mechanisms and skin factor in tight gas reservoirs. APPEA J

Chapter 5
Conclusions

The objectives of this study are to review, understand and evaluate the factors that have significant influence on formation damage and well productivity in tight gas sand reservoirs. Various tight gas field completion, production and reservoir engineering data are studied, laboratory experiments are performed, and reservoir simulation models are run. Based on the study results, the following conclusions can be drawn:

- Liquid phase trapping damage is one of the main factors that can significantly affect well productivity in tight gas reservoirs. The damage mechanism is mainly controlled by capillary pressure and its resulting relative permeability curves.
- Core flooding experiments using unsteady state technique integrated with numerical simulation approach, provided the relative permeability curves for a typical West Australian tight gas reservoir.
- Damage caused by water blocking is considered to be one of the main causes of low well productivity in water sensitive tight gas reservoirs (i.e. sandstones containing clays, especially smectite).
- In tight reservoir rocks that are sensitive to water, exposure of formation to water during drilling may cause severe damage to near wellbore formation permeability.
- Core flooding experiments using core samples taken from the typical West Australian tight gas reservoir showed that the damage to formation might be reduced by use of oil based mud drilling fluid, instead of water based mud.
- The field study and reservoir simulation highlighted that hydraulic fracturing may be inefficient in tight gas formations that are sensitive to liquid invasion damage. In some cases, hydraulic fracturing may even reduce well productivity.
- Underbalanced drilling is considered as a method to reduce damage to tight formations. However the study highlighted that even in the case of underbalanced drilling, mud filtrate (water or oil) may still invade the rock matrix around the wellbore due to the strong capillary pressure suction.

N. Bahrami, *Evaluating Factors Controlling Damage and Productivity in Tight Gas Reservoirs*, Springer Theses, DOI: 10.1007/978-3-319-02481-3_5,

- After the liquid leak-off into formation is stopped, during clean-up period when gas is produced from the tight reservoir, water still continues invading more area of the reservoir due to the very strong capillary pressure suction effect.
- Liquid loading in wellbore may significantly affect well productivity in tight formations, since the initial gas production rates are normally not high enough to lift the wellbore liquid to surface.
- In the case of damage to natural fractures, well productivity is reduced significantly. Well productivity of a damaged naturally fractured reservoir may not be very different than well productivity in a non-fractured reservoir until the reservoir contact area is very great.
- Field observations indicated that wellbore instability can cause large wellbore breakouts and washouts across the tight sand intervals.
- Cased-hole perforated completion may not be the optimum option in tight gas reservoirs. In cased-hole completed wells, the large wellbore breakouts filled by cement may reduce accessibility to the formation via perforation tunnels, as they may fail to penetrate deep enough into the formation. In addition, perforation jet penetration depths are significantly reduced by the tightness of the formation.
- Open-hole completion in tight gas reservoirs, provides using the advantage of enlarged wellbore (reduced skin factor) caused by large wellbore breakouts. Open-hole perforating using deep penetrating perforation charges may further improve the well productivity.
- In the case of perforation, the optimum option might be to use deep penetrating perforation charges oriented towards maximum stress direction i.e., 180° phasing (the direction where wellbore is stable and has no considerable breakouts).
- Use of the second derivative of transient pressure in welltest analysis can provide more reliable determination of reservoir permeability, skin factor and hydraulic fracture parameters in tight gas reservoirs.
- Integrating the diffusivity equation solution for linear and elliptical flow regimes in fractured tight gas wells can provide reliable estimation of reservoir permeability and hydraulic fracture size.
- In tight reservoirs with significant permeability anisotropy, drilling perpendicular to the maximum permeability provides higher productivity.
- If stress anisotropy is the main cause of permeability anisotropy, drilling long deviated/horizontal wells perpendicular to the maximum horizontal stress direction can result in achieving higher gas production rates, by intersecting the higher permeability conduits.
- Drill directional wells perpendicular to the maximum horizontal stress azimuth to improve wellbore stability, and intersecting higher permeability conduits, especially for the reservoir with normal faulting stress regime condition.
- In the case of fully perforated horizontal well, a longitudinal hydraulic fracture provides noticeably a higher productivity than a transverse hydraulic fracture.
- Welltest permeability and image log fracture aperture and fracture spacing can provide approximation of natural fractures permeability in naturally fractured tight gas reservoir, and result in more reliable evaluation of well productivity.

- Liquid invasion into tight formations may affect the formation pressure measurements and result in under-estimating reservoir pressure and over-estimating pressure gradient.

This thesis demonstrates how different well and reservoir parameters control well productivity and damage mechanisms in tight gas reservoirs. I have quantitatively shown the effect of phase trapping damage, well parameters and reservoir characteristics on tight gas sand reservoirs productivity for different types of wells and reservoirs such as non-fractured and hydraulically fractured wells, in under-balanced and over-balanced drilling, and in the case of invasion of different fluids into the tight formations. Reservoir simulation model for a typical tight gas reservoir is built and run, and analytical and numerical simulation approaches are integrated with core flooding experiments and field data analysis in order to characterize the key reservoir parameters and understand the effects of different parameters on well productivity. I determined the relative permeability curves for Whicher Range tight gas reservoir using core flooding experiments data analysis. I have quantitatively shown how the phase trapping damage can be reduced by use of oil based drilling fluid instead of water based fluid. I have introduced new techniques of well test analysis for tight gas reservoirs that can reduce uncertainties in estimation of average reservoir permeability, and also developed a new correlation that can determine permeability of the natural fractures in tight formations. I studied and analysed the different well completion, production and reservoir data from Whicher Range tight gas field in order to identify why production rates are significantly lower than expectations, and investigate possible remedial strategies to achieve viable gas production rates. Based on the study outcomes, I have proposed the optimum tight gas development strategies to achieve an improved productivity.

The manufacturer's authorised representative in the EU is Springer Nature Customer Service Centre GmbH, Europaplatz 3, 69115 Heidelberg, Germany. If you have any concerns regarding our products, please contact ProductSafety@springernature.com

Printed and bound by CPI Group (UK) Ltd, Croydon, CR0 4YY
31/07/2026
02182831-0002